Praise for

THE FENCE

"This could be a case study in the perils of profiling. It isn't so much forgotten history, as buried. Perhaps this outstanding book can fix that."   —Adrian Walker, *Boston Globe*

"Dick Lehr gets inside the heads of cops, criminals, prosecutors and politics better than anyone I know. *The Fence* is a revealing exposé of the blue wall of silence that endangers us all."

—Alan Dershowitz, Harvard Law School

"Dick Lehr has written a long overdue assessment of the brutal attack by law enforcement officers on an African-American plain-clothes officer, Michael Cox, which illuminates a tragic example of the failure of our criminal justice system. You will not be able to put this down and hopefully it will compel you to make sure that incidents like this never happen again on our watch. Every police officer, judge, lawyer and citizen should read this book."

—Charles J. Ogletree, Jr., Harvard Law School, author of *When Law Fails*

"Jolting, nightmarish and potent, this true cop yarn bests any bogus reality show or overblown tabloid tale with its hardboiled spin."   —*Publishers Weekly* (starred review)

"Like cancers that never seem to be cured, the inextricably linked ills of racism, public corruption and police misconduct continue to surface in Boston (and indeed in America). In his disturbing new book, Dick Lehr vividly presents another example of how difficult it is to face up to, let alone resolve, these conflicts."

—Former Los Angeles chief of police Bill Bratton

"Think Serpico translated to Boston. . . . A cautionary tale about the abuse of power and a timely civics lesson on the virtue of standing up to authority." —*Kirkus Reviews*

"Weaving mini-biographies of key players into the unrelenting police cover-up and Cox's determined quest for justice, he paints a gritty portrait of urban life that reads like a crime novel—replete with plot twists and vivid, deeply flawed characters. . . . The author's solid grounding in the history and diverse cultures of Boston, which he has covered for twenty-five years, adds context to the unfolding events." —Dave Holahan, *Hartford Courant*

"Intriguing . . . an admirable, in-depth description of police corruption." —*Library Journal*

"Gripping. . . . Tackles a broad issue with the zeal of a seasoned investigative newspaper reporter." —Lynet Holloway, *Black Voices*

"One of the darkest chapters in Boston police history is recounted both heartbreakingly and luminously." —*Rhode Island Lawyers Weekly*

Karin Lehr

ABOUT THE AUTHOR

A professor of journalism at Boston University, DICK LEHR was a reporter for nearly two decades for the *Boston Globe*, where he won numerous journalism awards and was a finalist for the Pulitzer Prize. He is the coauthor of the Edgar Award–winning *Black Mass*, the Edgar Award finalist *Judgment Ridge*, and *The Underboss*. While completing *The Fence*, he was a visiting journalist at the Schuster Institute for Investigative Journalism at Brandeis University. He lives near Boston, Massachusetts, with his wife and four children.

THE FENCE

ALSO BY DICK LEHR

THE UNDERBOSS:
The Rise and Fall of a Mafia Family
(WITH GERARD O'NEILL)

BLACK MASS:
The True Story of an Unholy Alliance Between the FBI and the Irish Mob
(WITH GERARD O'NEILL)

JUDGMENT RIDGE:
The True Story Behind the Dartmouth Murders
(WITH MITCHELL ZUCKOFF)

THE FENCE

A Police Cover-up Along Boston's Racial Divide

Dick Lehr

HARPER

NEW YORK • LONDON • TORONTO • SYDNEY

HARPER

A hardcover edition of this book was published in 2009 by Harper, an imprint of HarperCollins Publishers.

Grateful acknowledgment for permission to reproduce images appearing in the insert is made to the following: The Bostonian Society: 1, top. Wooster School: 1, center; 3, top. Kristine Conley Cox: 1, bottom; 2, bottom. Mattie Brown: 2, top left and top right; 3, bottom. The Suffolk County District Attorney's Office: 4, top and bottom; 5, top. The *Boston Globe:* 5, bottom; 6 top and bottom; 7, top, center, and bottom; 8, top. The author: 8, bottom.

HarperCollins books may be purchased for educational, business, or sales promotional use. For information please write: Special Markets Department, HarperCollins Publishers, 10 East 53rd Street, New York, NY 10022.

First Harper paperback published 2010.

Designed by Eric Butler

Map of Boston © David M. Butler

Library of Congress Cataloging-in-Publication Data is available upon request.

ISBN 978-0-06-078099-9

10 11 12 13 14   OV/RRD   10 9 8 7 6 5 4 3 2 1

TO KARIN
FOR EVERYTHING

HUNGER ONLY FOR A TASTE OF JUSTICE

HUNGER ONLY FOR A WORLD OF TRUTH.

—Tracy Chapman,
"All That You Have Is Your Soul"

CONTENTS

THE **CAST OF CHARACTERS**

ROBERT "SMUT" BROWN	*Mattapan drug dealer and shooting suspect*
MATTIE BROWN	*Smut's mother*
JIMMY BURGIO	*Boston police officer assigned to Dorchester*
DONALD CAISEY	*Boston police officer in the anti-gang unit*
KENNY CONLEY	*Boston police officer assigned to the South End*
MIKE COX	*Boston police officer in the anti-gang unit*
KIMBERLY COX	*Mike's wife*
IAN DALEY	*Boston police officer assigned to Roxbury*
WILLIE DAVIS	*Conley's attorney*
SERGEANT DAN DOVIDIO	*Burgio and Williams's supervisor*
BOBBY DWAN	*Conley's partner*
JIMMY "MARQUIS" EVANS	*shooting suspect and Tiny's brother*
JOHN "TINY" EVANS	*drug dealer and shooting suspect*
PAUL EVANS	*Boston police commissioner*
JIM HUSSEY	*head of Boston police Internal Affairs*
LYLE JACKSON	*shooting victim at Walaikum's*
CRAIG JONES	*Cox's partner*
TED MERRITT	*federal prosecutor*
SERGEANT DAVID MURPHY	*supervisor at Woodruff Way*
BOB PEABODY	*assistant Suffolk County district attorney*
INDIRA PIERCE	*Smut's girlfriend*
JIMMY RATTIGAN	*Boston police officer assigned to Roxbury*

THE **NEIGHBORHOODS**

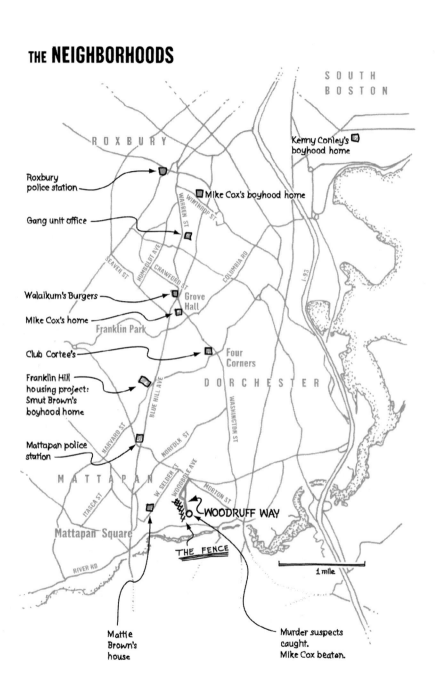

SOUTH BOSTON

ROXBURY

Kenny Conley's boyhood home

Roxbury police station

Mike Cox's boyhood home

WARREN ST
WINTHROP ST

Gang unit office

SEAVER ST
HUMBOLDT AVE
CRAWFORD ST
COLUMBIA RD
I-93

Walaikum's Burgers

Grove Hall

Mike Cox's home

Franklin Park

Club Cortee's

Four Corners

DORCHESTER

BLUE HILL AVE
WASHINGTON ST

Franklin Hill housing project: Smut Brown's boyhood home

Mattapan police station

HARVARD ST
NORFOLK ST

MATTAPAN

ITASCA ST
W. SELDEN ST
WOODBOLE AVE
MORTON ST

WOODRUFF WAY

Mattapan Square

THE FENCE

RIVER RD

1 mile

Mattie Brown's house

Murder suspects caught. Mike Cox beaten.

THE FENCE

Prologue: January 25, 1995

When Kimberly Cox was awakened by the telephone ringing in the middle of the night, the fourth-year medical student had been sleeping hard. She'd slept through the Boston police and ambulance sirens blaring an hour earlier on Blue Hill Avenue two blocks from her home. She was likely used to the discordant sounds; the wail of sirens was not unfamiliar in Dorchester, where she and her family lived, one of the many black families making up the neighborhood.

When the phone rang, she was alone in bed. Her first thought was that her husband, Michael, was calling. Michael was usually home by 2 A.M.; if he was going to be later he would call. Then Kimberly noticed the clock: It read 3:30.

She picked up the receiver.

Mrs. Cox?

Kimberly did not recognize the voice.

The caller identified himself as Joe Teahan, an officer with the Boston Police Department. Kimberly worked to clear her head. The name meant nothing to her. In fact, Teahan was a white officer who worked with her husband in the department's elite anti-gang unit, composed of officers working primarily in street clothes, who targeted the street gangs of Roxbury and Dorchester. The gang unit's supervisor had instructed Teahan to call Mike's wife. "Just don't scare her," the sergeant had said.

Kimberly listened as the voice told her Mike had been in an "accident."

What kind of accident?

He's alive but hurt. He's on his way to Boston City Hospital.

Kimberly was up and standing by the bed. She was nervous all over. She dressed quickly. Teahan said they would send a car to get her. But it wasn't that simple. Fast asleep in their bedrooms were her boys, six-year-old Mike Jr. and Nick, whose fifth birthday was still fresh on everyone's mind. She told Teahan she'd call him back after figuring out the logistics. She hung up and hurriedly dialed her mother-in-law. Kimberly was thinking Bertha Cox could stay with her sons. But when Bertha arrived a few minutes later, she insisted on going along with Kimberly to the hospital. This led to more telephone calls to other family members to ask them to hurry to 52 Supple Road, where Michael and Kimberly and their boys lived in the second-floor apartment of the two-family home owned by one of Michael's sisters.

It took nearly an hour for Kimberly to sort it all out. That's when Joe Teahan and his partner, Gary Ryan, pulled up in front of the red-brick house. The two officers had first met as classmates in the police academy and had worked side-by-side for most of their four years on the force. They could see that fellow officer Mike Cox was living in the middle of it all. Walking a block in any direction landed you on a street where guns and drugs were the name of the game. In fact, for Mike and the dozen or so gang unit officers on a special operation that night, the trouble had begun only three blocks away.

Kimberly and Bertha Cox came down the brick front steps, hurried across the cement walkway of the tiny front yard, and climbed into the backseat of the cruiser. The two officers stuck to the script. They told Kimberly that Mike had head injuries. They said he likely slipped on a patch of ice, hit his head, and "split it open pretty good." Gary Ryan did not share what he'd thought when he first saw Mike on the ground, his head so bloodied and swollen, "it looked like a gunshot wound." The sergeant had said not to alarm her. Little else was said during the short drive to the hospital about a mile away.

The two women were taken to the emergency room entrance. They rushed through a double set of automatic doors. The entryway's linoleum floor was covered with a carpet rolled out in wintertime to absorb wet snow and slush. Nurses steered them through the trauma unit's two heavy wooden doors that swung outward.

For a weeknight in the winter, the emergency room at Boston City Hospital was a busy place. Surgeons and nurses in the operating room were working furiously on a man named Lyle Jackson. Jackson, twenty-two years old, had been shot three times in the chest by two gunmen at a small take-out restaurant on Blue Hill Avenue, where he'd gone to munch on chicken wings and a hamburger. The young Roxbury man had been in the ER less than an hour. Meanwhile in the acute care unit, two other Boston police officers were receiving treatment. Jimmy Rattigan occupied one of the thirteen bays, and his partner, Mark Freire, was in another. Both had been injured while chasing Lyle Jackson's shooters. Their cruiser was demolished when it hit a parked van on a narrow Roxbury street.

Kimberly and her mother-in-law found Mike in one of the other bays that circled the unit, each enclosed by a cloth curtain hanging from the low ceiling. When she first saw her husband, Kimberly said nothing. She walked up to where he lay on a gurney and studied him. It was as if in these surroundings, Kimberly, anxious up to this point, switched gears and assumed the detachment of the budding physician that she was. Observations she made were clinical: hematoma, the size of an egg, on the patient's head; a swollen face; swollen nose; one laceration on his scalp that would require sutures, another on his lip. Multiple abrasions and scratches.

Kimberly Cox was confused. She was looking at someone suffering from multiple injuries—serious injuries. She thought, "It didn't look like he had slipped and fell."

When Boston police officer Craig Jones stepped through the curtain and into the tiny ten-foot by twelve-foot bay, Kimberly recognized him right away. Craig was Mike's partner and good friend. Kimberly saw that Craig was agitated, even upset. There were other officers with him whom Kimberly did not know. One or more of them crowded inside too. Kimberly and Bertha Cox were seated by Michael's side.

Then another police officer, a black officer, tall and dressed in uniform, appeared at the curtain. He did not actually step inside, but poked his head into the bay. The officer was addressing Craig Jones, but Kimberly heard him say, "I think I know who did this."

Craig Jones and the others stepped outside. To Kimberly, the talk was mostly in covered, hushed tones. But this she was able to hear: "I think cops did this."

Kimberly was speechless. Bertha Cox was not.

"What?" she asked. "Police officers did this?"

The question hung in the air.

"Oh my God," Bertha Cox said.

By dawn, the morning TV news stations in Boston were broadcasting reports about the shooting of Lyle Jackson, the high-speed chase that followed, and the dramatic capture of four men suspected in the shooting. The city's leading newspaper, the *Boston Globe*, also ran a story in its morning edition about a shooting that turned into a homicide case when Lyle Jackson was pronounced dead at Boston City Hospital. The newspaper stories mentioned officers Rattigan and Freire and the injuries they had sustained.

But no story mentioned anything about Michael Cox.

PART I

Two Cops and a Drug Dealer

CHAPTER 1

Mike Cox

Boston police officer Mike Cox directed his partner to swing by his house before heading over to check out the scene at the club Cortee's. The apartment at 52 Supple Road in Dorchester was only about a mile from the club. Mike ran inside and pulled off his black nylon Windbreaker, the one with his unit's patch stitched on the left breast beneath the Boston PD emblem. He changed and hustled back outside.

It was just before midnight. The below-freezing weather wasn't the reason for the clothing switch. Mike needed to fit in, and within minutes he and his partner were walking into the low-slung building with its unwelcoming dirty brick exterior. The club resembled a warehouse about to be condemned. The three windows in front were so narrow and smeared that inside they guaranteed darkness, not light. The only touch of style was the rooftop sign— with the name, Cortee's, written in a swirling red script across an orange background.

The club was smack in the middle of a neighborhood known as Four Corners, which was targeted periodically by city leaders and neighborhood advocates for urban renewal. The newspaper article announcing one such effort described Four Corners as "a neighborhood where mothers do not let young children play in front yards.

Where nearly 40 percent of families with children under age five live in poverty. Where teenagers keep their eyes open and routinely throw furtive looks over their shoulders. Where empty lots and 'for sale' signs scar almost every block. Where street justice is the law of the land."

At the club on a Saturday night two months earlier, a near-riot had broken out; a young Dorchester man was stabbed in the butt and the back and taken by ambulance to Boston City Hospital. When police arrived they found a crowd outside shoving and fighting and throwing bottles. Three men were arrested in connection with the stabbing, and police caught two other men slashing the tires of a cruiser.

Mike Cox squeezed past patrons to make his way deeper inside. Next to him was Craig Jones. Right away they liked what they saw: The club was running at full throttle, the music blaring across a large room jammed with at least a couple hundred people. Hip-Hop Night guaranteed a crowd, even on a weeknight in January when the temperature was 29 degrees and Boston's inner-city streets were mostly vacant.

No one noticed them as cops. Craig was dressed in blue jeans and a black-hooded sweatshirt underneath a black leather jacket, while Mike wore a three-quarter-length black coat—the kind of puffy, hooded, goose-down parka popular in the 'hood. Underneath he had on black jeans and a black sweatshirt. A black Oakland Raiders wool hat was pulled over his head. Mike had assembled the outfit for assignments like this, borrowing the parka and skullcap from a teenage nephew.

Mike's getup was the more elaborate of the two, which was no surprise. He was clothes-conscious—and always had been. Craig had certainly noticed this during the five years they'd been partners. There was another cop in their unit who sold jackets, sweatshirts, T-shirts, and hats featuring Boston police patches and logos. Mike was a regular customer. The Windbreaker he wore most nights while on patrol had come from this guy. "Mike liked those police clothes," Craig said.

They walked to one side of the club to sit in a couple of empty chairs. Together they made quite an impression. Mike was six feet, two inches tall and weighed 215 pounds or so, and Craig was even taller, by two inches. They were both strong and fit and athletic. Mike's parka was long enough to conceal his handcuffs, badge, and semiautomatic handgun, a gun smaller than the one issued by the department that fit snugly into a tiny holster.

The two were members of the department's elite Anti–Gang Violence Unit, or AGVU, a collection of forty cops who roamed the meanest streets of the city in pursuit of street gangs, drugs, and guns. "You had more freedom to investigate much more serious crime," Craig once said about why he found satisfaction in the role as street-gang fighter.

Their call sign, for police radio purposes, was Tango K–8, or TK–8.

"Tango" stood for the gang unit.

"K" meant they worked in plainclothes.

"8" stood for them—Cox and Jones.

By 1995, Mike had been on the force for six years, Craig for eight. Mike was twenty-nine and Craig was thirty. Through work, they'd become close friends. Craig brought his daughter to Mike's house for his son's birthday party, and Mike went to Craig's house for his daughter's party. Sometimes they'd go out together and shoot pool, and, for a bit, they played basketball on the same team in the police league. In fact, Craig was probably the only one in the gang unit who knew anything about Mike's background and personal life—that Mike, for example, had grown up in Roxbury a few blocks from the gang unit's offices at 364 Warren Street. Or that Mike had attended a private high school in western Connecticut. Mike was especially sensitive about that. On a police force where the officers were mostly working class, white, and mostly educated in urban high schools, no way he wanted to be seen as a black preppie. Indeed, when it came to his personal life, Mike went mute.

They worked as a team in plainclothes. The public often confused working in plainclothes and working undercover. The two were vastly different. Undercover meant assuming a phony identity to infiltrate a criminal organization. Plainclothes work meant simply not wearing a uniform on the job. It also meant driving an unmarked car—a police vehicle without the blue bubble on the roof and the blue-and-white coloring and lettering on the exterior. Mike and Craig drove a dark blue Ford with no blatant police markings. But it was equipped with a siren; blue lights concealed in the front grille; wig-wag, or blinking lights, in the rear; and a blue light on the front dashboard.

By blending into the street, Mike and Craig were looking for an edge. It was unrealistic to think street-smart gang members would not spot them or their unmarked car. But what they were looking for was a few extra beats before the click of recognition. "It helped me, you know, you'd be right on the scene, or very close to someone before they recognized you as a cop," Mike said. The gang unit cops valued those extra seconds, whether during a routine patrol of a housing project or during a raid.

Mike and Craig were inside the Cortee's to perform some quick reconnaissance near the end of their regular shift. It was part of a larger plan. To cops, Hip-Hop Night might as well have been called Gang Night. The music attracted the gangs, and, Mike said, "wherever they go, there is going to be trouble." He and Craig had gotten "intelligence," or word from informants, there was conflict in the air. Mike and Craig wanted to size up the scene and make certain the club was humming. They and a handful of other teams from the gang unit then planned to return and set up outside before closing time—2 A.M.—when the crowd would begin pouring into the street.

The unit had high hopes for a rich return on their investment.

Seated at the table, Mike and Craig looked around. Even if they'd wanted to talk, the music was too loud to be heard. The only light was over by the bar area. Neither was too worried that anyone would make them. "They'd really have to get right in your face to recognize you," Mike said. But it was their look that was mostly the source of confidence in not being identified. "Craig is a good-sized guy, and I'm a good-sized guy, and we're dressed up like that. Most people are *not* thinking, Oh, cops.

"They're thinking, Whoa! Like, stay away from them."

Mike saw gangbangers from a number of different street gangs—Humboldt, Castlegate, and Corbett, to name a few. "Craig would see somebody, hit me, and I'd look, and I'm like, yeah, yeah, I see." They both felt the combustibility in the room. The two stood up, walked back across the floor, and left. They'd seen enough and would be back, confident closing time would be the right time for the gang unit to be hiding nearby in the dark.

The civil rights movement made itself heard in Boston in 1965 with a legal blockbuster. The Boston branch of the NAACP sued the city's school committee in federal court seeking to desegregate public schools. The lawsuit marked the formal beginning of legal and civil conflict, building nine years later to court-ordered busing.

Two days after the court filing, Dr. Martin Luther King Jr. flew into Logan Airport to tour Boston and then lead an estimated twenty-two thousand protesters from a playground in Roxbury to the historic Boston Common downtown. In the rain on April 23, King said, "The vision of the New Boston must extend into the heart of Roxbury. Boston must become a testing ground for the ideals of freedom." The crowd roared with approval.

Two months later, on June 17, 1965, Michael Anthony Cox was born to a family living in the heart of Roxbury. He was the youngest of six children. His parents, Bertha and David, had lived in Boston for a decade. They were from Tennessee, where they'd met, married, and started a family. Their move north was part of the great migration of blacks during the 1940s and 1950s. They had followed Bertha's mother, Rosa, the first to come, who found work as a maid for a wealthy Jewish couple, Norman and Helene Cahners. Norman Cahners oversaw a publishing empire he built from a single magazine, and he was known for his generosity and philanthropy.

They owned homes in the city and in Brookline, just west of Boston, and Mike's grandmother Rosa moved between the different homes. The Cahnerses also purchased a house in Roxbury for her to live in: 62 Winthrop Street, which was half of a large, side-by-side two-family home.

When Rosa's Jewish neighbors put their side on the market, Rosa urged her daughter Bertha and son-in-law to move north, and on January 20, 1955, they bought 60 Winthrop Street. The Coxes paid $6,500, borrowing $5,500 from a Roxbury bank. Everything fell into place nicely. Bertha and David settled into 60 Winthrop, with Rosa right next door. Rosa then persuaded another daughter, Ollie Parks, and her husband to move north; eventually Ollie went to work for the Cahnerses as well.

The Coxes' Winthrop Street was a one-way street running westerly from Blue Hill Avenue, Roxbury's main thoroughfare. From the opening at Blue Hill Avenue, the street consisted mostly of small apartment buildings and homes, many in disrepair. The Coxes' house at 60 Winthrop was toward the other end of the street, a couple of blocks from Dudley Square. The buildings were better kept on this end. Even so, coming upon 60 Winthrop required a double take. The structure was oversized, even for a side-by-side two-family, with the Coxes' number 60 sharing a center wall with its mirror image at 62 Winthrop Street. But more distinctive than its size was its unusual architectural style. "One of the more robust manifestations of Italianate style in Roxbury and the Boston area," noted the city's Landmarks Commission.

The Coxes occupied a home that reflected the full arc of Roxbury's social and ethnic history—from the original Puritan settlers to Irish, Jewish, and then African American. The land was originally owned by the Reverend Thomas Weld, who, along with his brothers, emigrated from England in the 1630s and came to own hundreds and hundreds of acres of land in the Massachusetts Bay Colony. The Reverend Weld became the first minister of the First Church in Roxbury in 1632, and the entire Brahmin family became deeply embedded in the state's history; in modern times, they included the actress Tuesday Weld and the state's sixty-eighth governor, William F. Weld, who served from 1991 to 1997.

In 1852, Samuel Weld built the large double Italianate on Winthrop Street. Its main entrance—located on the side—had gabled door hoods and an oriel, or bay window, above it. The large pine front door opened to a spacious entry featuring high ceilings and an elliptical staircase curling upward to the second floor. It was the kind of splashy, grand entrance showing off the owners' standing and wherewithal.

For the remainder of the century and into the 1900s, the house was owned by Charles D. Swain. Swain was a rich man, a prominent merchant who owned one of the largest stores in the bustling and fast-growing Dudley Square nearby. The Swains and later owners of 60 Winthrop Street were insulated from any development by abutters when an order of Carmelite nuns moved next door in the 1890s. The order built a monastery and enclosed the grounds behind a brick wall fifteen feet tall.

The house changed hands, just as the neighborhood did, with the Yankees giving way to the Irish at the turn of the century. In 1914, when he began serving his first term as mayor, the legendary and charismatic James Michael Curley lived one street over on Mount Pleasant Avenue. By the 1920s, the Irish were moving out of Roxbury, replaced by Jews, and during World War II the neighborhoods of Roxbury, Dorchester, and Mattapan—a three-square-mile area—became home to ninety thousand Jews. Blue Hill Avenue, observed the historian Thomas H. O'Connor, was "often derisively dubbed 'Jew Hill Avenue' by members of other ethnic groups." Following the war, the Jews migrated south into the suburbs, and Roxbury became a black neighborhood.

Many blacks moving to Boston filled the housing projects sprouting up all over Roxbury during the postwar building boom. Owning a home was less typical—and put Mike's parents squarely in an emerging black middle class. The climb up the economic ladder was nothing less than hard-earned. Mike's father was known for his work ethic; he went to work as a boy, ending his formal education after the sixth grade. In Boston, he was the first black to own a landscaping business, D. E. Cox Landscaping. He also eventually owned florist stands, one in Dudley Square and one downtown. He was a heavy-smoking man, lean and small, known for his quietness and long hours on the job. In time, Mike's mother went to work at Raytheon, where she was a wire sorter at the defense technology company in Waltham for nearly three decades. One neighbor said Bertha and David Cox "worked very, very hard to make a better life for their kids."

When Mike was born he was truly the baby of the family. The three sisters, Cora, Lillian, and Barbara, all born in Tennessee, were in their mid-to-late teens. His brother David was fourteen, and Ricky was seven. Mike was surrounded by women who not only looked after him but told him what to do: his mom, grandmother, and sisters. "I was talked *at*, I wasn't really talked to," Mike said. His father, meanwhile, displayed the same firm hand over family affairs as he did with the landscaping business. "He had them all in line," a neighbor said.

Under a careful eye, Mike was allowed to play at a nearby park. His parents mostly made him stick close to home, where he'd shoot basketball at the hoop in their driveway. Mike had no idea about his home's fancy bones. The chandelier in the entry was long gone, and the handsome wood floors were covered with wall-to-wall carpet. Even if he noticed it, he never wondered about the remnant wiring still strung along the creases in the walls or ceilings—wiring for the "call bells" the Swains used to summon servants from their quarters on the third floor. The sinks in the bedrooms were left over from when servants carried water to the rooms in buckets because there was no running water. Mike viewed the large house with its three floors and two staircases as a playground.

His bedroom as a boy was the tiny room above the front entry with a country view, completely disconnected from typical Roxbury. The room looked over the high brick walls enclosing the Carmelite monastery: green lawns dotted with maple, oak, and fir trees. In springtime, the apple and cherry trees blossomed pink and white flowers. Bells rang daily, calling people to prayer and, as the nuns would say, "directing their thoughts to the faithful presence of God in their midst."

The house at 60 Winthrop Street was like a sanctuary, sequestering Mike from the trouble not far away. During the 1960s and 1970s, the Dudley Square area became a no-man's-land. "A day doesn't go by without a stabbing or shooting or assault with a baseball bat or club," a judge in the Roxbury District Court once said. "Unemployment is bad. Housing is bad. The schools are bad. When you have these conditions, you're going to have crime." Merchants suffered, and many stores closed and storefronts were boarded up. "People are afraid to walk the street," complained one merchant to the *Boston Globe*. In 1971, when Mike was five, the city opened a new, $4 million police station across from where Winthrop Street ended at Dudley Square. More than three hundred officers were deployed. Several months later, a new $3.5 million glass and concrete courthouse was built next door. Both were seen as necessary to tackle the sharp rise in murders, rapes, and drugs.

Mike got to know some of the officers, and they seemed nice. He was also drawn to shows on TV about police, and he daydreamed about being a police officer. But the idea felt daring and even intimidated him. "I didn't really think I could be a police officer for some reason. I didn't know if I was good enough." He really liked one officer named Will Saunders, a family friend who made a lasting impression on Mike, mainly because of his race. "There were not very many minority officers then, so he really stood out."

Father and son shared a number of traits, but most notable was their reserve. In kindergarten, Mike suffered one of his recurring nosebleeds during nap time. He didn't say anything to the teacher, and just lay there as the blood pooled. When nap time ended, the teacher saw Mike and was aghast. His parents were called in for a meeting. "He doesn't talk," Mike recalled the teacher saying with worry. "All he had to do was come get us and he didn't even do anything."

Cauterizing resolved the nosebleeds, but Mike's quietness continued. It was a reason he repeated the first grade. "I was immature," said Mike. "I didn't talk. That was a big thing." For elementary school, Mike followed his older brothers to a private Catholic parochial school in nearby Brookline, St. Mary's School. "My folks didn't know much about education, you know, my dad had a sixth-grade education and my mom, I think, eleventh grade. But they knew the Boston public school system wasn't that good." In the early 1970s, the escalating battle over forced busing was in the news all the time, and his parents grew alarmed. The Coxes were neither particularly political nor religious; they just wanted something better for Mike and his siblings—so, before there was "white flight," the name given to waves of white Bostonians fleeing the city school system, there was "Cox flight." Mike's grandmother Rosa heard about the Brookline school from her employers, and, Mike said, "She passed that tip on to my mother and father."

For his parents, the school was a stretch financially—a couple of hundred dollars in tuition plus the cost of school uniforms. He sometimes heard "grumblings" from his father when the bills were due. For his part, Mike just followed along, even if privately he wondered why he had to attend a school so far from home. It seemed so far away because of the family's early morning routine—Mike was out of the house by 6 A.M. to ride along in the station wagon while his father drove his mother to work in Waltham and then backtracked to drop him off at school in Brookline. Mike took the bus home from school and went next door to stay with his grandmother. "My mother wouldn't get home until later, and my dad worked pretty late all the time."

St. Mary's had about two hundred students. There was one class for each grade, with fifteen to twenty-five students. Mike realized right away he stood out—he was usually the only black student in his class. But he did get used to his surroundings. "There were a lot of kids who, although we looked different, we had a very similar background. Their parents weren't wealthy. They were hardworking, middle-class people." Nonetheless, things happened to remind him he was different from most kids at the school.

It happened once when he was eight when his aunt Ollie landed in the spotlight. Working for the Cahnerses, she answered the front door on January 19, 1974, to find a red-haired woman on the stoop. The Cahnerses were in Florida on vacation. The woman began asking for directions and suddenly pulled a pistol from her coat pocket. At the same time, a man wearing a ski mask stepped into view and pointed a gun at her. The burglars taped Ollie's hands together, made her sit in the foyer, and stuffed paper into her mouth. They raced from room to room, yanking paintings off of walls. They fled with three, including *The Rustics*, by Winslow Homer, valued at up to $200,000. The next day's newspaper coverage was extensive. "Masked Pair Loot Brookline Home of Publishing Executive," ran the *Boston Globe* headline over a story that recounted how Ollie freed herself after the "bandits" left.

In school the next day the armed burglary was a hot topic, and one of Mike's classmates carried a copy of the newspaper, which included a photo of Ollie.

"That's my aunt," Mike said.

Your aunt's a maid?

Mike was embarrassed. He said no more and realized he should have kept his mouth shut.

By the time Mike was in the seventh and eighth grades—spent at a middle school in the city—teachers were encouraging him to spread his wings. Mike began reading a lot, thanks to an English teacher. "She'd hand me a bag full of books—read these!" His grades were strong and he was a natural athlete. With his teachers' guidance, he applied to several private schools. Milton Academy offered him a scholarship. Mike liked the school because it was fairly close to home, in the town of Milton south of the city.

In September 1980, Mike began the ninth grade at the elite

private school. His father drove him, but the school year was barely under way when David Cox fell ill. He was diagnosed with stomach cancer. Mike recalled, "He had surgery and they took out part of his stomach, and he was pretty ill. He came home and had lost of a lot of weight, a lot of weight, and he had stopped working." The family was in crisis.

For Mike, Milton Academy was a crisis—academically and culturally. To get there Mike began traveling a network of buses and trains. He invariably arrived late. He usually got a ride home after playing sports, but it was well after seven o'clock before he could even think about his schoolwork. "But I am flat-out tired, and I'd go upstairs and go to sleep." If he was lucky, he'd wake up early to do some work. "If I didn't wake up I'd go to class and now I haven't done the work. So I'd try to do it between periods."

His head spun, but given his nature he said nothing. He didn't ask for help at school. He didn't say anything to his parents. "I didn't want to disappoint my father." He knew his father had cancer, but wasn't sure what that meant.

"No one really explained it to me, but I could see he was getting sicker."

Because of the complicated commute, Mike often missed meeting with his adviser before classes—meetings that were part of the fabric of the academy's day. One day his adviser caught up with him. He pulled Mike aside. Mike rubbed his eyes and sneezed. He'd begun suffering from allergies, although the condition hadn't yet been diagnosed. Mike just knew his head was stuffy all the time and his eyes watered constantly. The adviser waited a second and then said he had a question to ask.

You smoke a lot of pot, don't you? Before school?

Mike was dumbfounded.

You can tell me, the adviser said earnestly. It's okay.

Mike sat there. To him, the world was divided into two groups—kids and grown-ups. With friends, he felt okay, and "I did what I did. Played sports and was friends." With adults, "I just didn't talk. Talking wasn't my thing." Facing his adviser, Mike basically didn't say a word. He did not speak up and protest, did not seize the opportunity to discuss his rough start. "I was just sitting there, thinking, I don't know what you're talking about."

The adviser took Mike's silence as confirmation. You really should stop, he said. You really should.

The meeting ended. Mike left feeling more disoriented than ever, and the feeling just worsened as the year went on. Most of all, he felt alone going through the biggest culture shock of his life, a shock that was not about race. He'd attended a largely white school at St. Mary's, so being the rare black at Milton was not a foreign experience. It was the wealth; he'd never been around or seen such wealth before. He became acutely self-conscious. Seeing some of his classmates' mansions left him paralyzed socially. "I was petrified to bring anyone from school to my house. It was just embarrassing, you know. Oh my God, look at the house I live in, look at how these people live." He dodged conversations on campus when classmates talked about where their fathers went to college—Yale, Princeton, Harvard, and other elite schools. He was embarrassed his aunts and grandmother were maids. He was even embarrassed his mom's name was Bertha.

Mike could not figure out how to make Milton work. He was a day student when most of the kids boarded on campus. With his commute, team sports, and the piles of homework, he found himself in a hole academically. "I had a lot of D's and C's at first." It wasn't as if the schoolwork was too difficult for him, he just could not find time to complete it. "The perception was that I wasn't doing my homework because I couldn't do it. But I wasn't doing my homework because, at the time, I was just tired all the time and, I mean, there was a lot of stuff going on in my life."

Meetings with teachers and advisers did not help—mainly, once again, because Mike let stand their assumption that the work was over his head.

You can't do this, can you? Mike was asked. It's really difficult, right?

"I was just like, 'Yeah, I guess so.'"

By early spring, Mike thought he was making progress. Not playing a sport, he had more time for his schoolwork, and his grades improved. But it was apparently not enough. "My adviser went from 'You smoke pot, don't you?' and 'You have trouble doing the work, don't you?' to 'You don't really want to be here, do you?'"

It became a refrain: You're not happy, are you?

Mike did nothing to rebut the school's wrongful assessment, a response that was becoming a pattern. The next thing he knew, he would *not* return to Milton Academy. His mother and oldest sister Cora began working with administrators at Milton to find Mike a new school. In the spring and summer he visited other campuses, such as the Northfield Mount Hermon School in western Massachusetts and the Wooster School in Danbury, Connecticut. Mike was not impressed. "They all looked the same to me," he said. And both schools meant he'd have to leave home and board. "I didn't want to be a boarder."

But what Mike thought did not matter. "My mother told me I was going to Wooster because it was a full scholarship and everything was booked—the whole nine yards," said Mike. "So it was decided for me."

In 1981 when Mike went to the Wooster School, he joined a sophomore class that numbered between thirty and forty students. The Episcopalian school, founded in 1926, took pride in its small size and its progressiveness. It became coeducational in 1970. Mike's class had a few more boys than girls in it, and he was one of a handful of blacks. One of the other black scholarship students—a senior during Mike's first year—was Tracy Chapman. They had very different interests. Mike's focus was sports; Tracy was interested in politics, the black feminist poet Nikki Giovanni, and her music. She played at the campus coffeehouse Friday nights, where she sang songs she was working on, including one she titled, "Talkin' 'bout a Revolution."

Mike's mom, dad, and sister drove him down from Boston at the start of school in September. Most of the other kids were already on campus and unpacked. Mike was assigned to a triple. His roommates had set up the room to their liking. "I got the top bunk, whatever cabinets were empty—the leftover stuff." Mike's sister, surveying the situation, made it clear she didn't like that her brother got the short end. "She's like making faces, saying, 'Oh, I can see you got the worst of this.'" Mike just wanted them to leave. His sister and mother shuffled between the car and the second-floor room. "I'm embarrassed. I got this teenage thing going on, you know, with a mother and sister running around and saying things to other kids. I'm like, 'Oh, God, please, these people don't even know me. Just leave.' And my dad is so sick he never got out of the car."

Escorting them to the parking lot, Mike made his farewells—as quickly as he possibly could. His father was in his seat, frail and shrunken. "He didn't say a lot—study hard, stay out of trouble, that kind of thing. I love you." Like his father, Mike had little to say in return. He gave his father a quick hug and said, "See you later, Dad."

It was the last time Mike saw his father alive. The fall semester began and Mike became quick friends with one of his roommates, a boy from New Jersey named Tim Fornero. Mike played football and adjusted well academically—nothing at all like Milton Academy. But Mike didn't like being away from Boston. "The school was fine, the people were fine, but my mind, you know, was not at that school." The major distraction was his dad's health. Mike called home most weekends and talked to his mother. But when he asked for his father, he was told his father was too sick to come to the phone.

"I didn't really understand it," Mike said. He was never given specifics about his father's condition. Then just before Thanksgiving break he was called into the administration offices, where one of the school officials broke the news: His father was dead. Mike went blank.

It took a few days—after he'd gone home early for the holiday break for the funeral—for the shock to transform into anger. He learned his father died at home. Except for an older brother living in Michigan, everyone else was at his bedside. "I was totally out of the loop." Mike was incredulous that his mother had not told him his father was dying and summoned him home. In response, Mike went deeper into his shell. He was angry at his family and at the world. He mostly kept to himself during the holidays, and when January came around and it was time to go back to Wooster, he refused.

The rebellion was uncharacteristic, and his mother would have none of it. She drove him to Connecticut, and, back at school, Mike shut down. "I don't want to be there." Friends and teachers tried being supportive, but Mike rebuffed them. He would say he was leaving after sophomore year, so why bother? "I wasn't mean or hostile, I was just like, I'm not going to be here." That summer his mother had a different idea, and, again, she prevailed. Mike did return for his junior year.

Mike continued to smolder, a potent mix of anger and grief. In a show of protest, he did not play wide receiver on the football team, and instead played goalie for the soccer team. With winter approaching, he told people he was not going to play basketball— his favorite sport—and was going out for wrestling instead. But then Mike felt something give way. "Somewhere around then I got over—whatever, some of those issues." The sky cleared. "I realized I have a lot of friends, and I'm actually kind of happy, because I really didn't have a beef with anyone and the people were very nice."

The notion of not playing basketball suddenly seemed crazy. "I loved basketball." Mike and a teammate named Vincent Johnson became the dominant players for the talented Wooster basketball squad. They had become close friends through the sport, often staying after practice to play one-on-one against each other. Vince was a scholarship student from Washington, D.C. In many ways, they were polar opposites—in terms of both their game and their personalities. Vince had never played much basketball before, and he relied on pure athleticism. "All I knew was to put the ball in the hoop," he said, "and I was just going to keep at it until I got it in."

Mike's game was polished from years of playing organized basketball in Boston. "Mike was real finesse," said Vince. "He could float through the air." Their senior class yearbook featured a photograph of Mike in mid-air, gliding smoothly toward the basket past two opponents en route to making a left-handed layup.

Vince's personality overflowed with self-confidence and he displayed a fierce competitive streak. "Even if I'd never played something before, I was going to learn and win." Mike, in contrast, was selfless and diplomatic. "Mike would never run up the score on you," Vince said.

Mike would also serve as Vince's peacemaker. During a pickup game sophomore year, the player Vince was guarding faked him by pretending to take a shot—a pump fake—and Vince jumped. He was airborne, waving his arms wildly, watching helplessly as the player dribbled around him and easily scored. Vince looked so foolish; a heckler from the sidelines yelled, You see that big bird fly! Vince raced over and was right in the boy's face. The boy happened to be a senior, a star of the football team. He pushed Vince away. Vince came back and hit him in the face. Fighting broke out, but Mike stepped in. He got them to stop by raising the race factor, pointing out the idiocy of two black students pounding on one another. "He was like, 'Don't screw up. That's what they want. There aren't many of us,'" Vince recalled.

Senior year, Mike and Vince were the team's cocaptains. Tim Fornero was the manager. They had a blast. The team went undefeated, piling up a 15–0 record and capturing the Hudson Valley League Championship. Mike was scoring twenty or more points a game. But their run ended abruptly in the state tournament, when they lost a playoff game that saw Mike hobbled for the craziest of reasons. He'd forgotten his sneakers, and had to borrow a pair. "His toes were like bleeding through the game," Vince said. "I think that's why we lost; he wasn't playing regular."

Mike was named one of the team's most valuable players. He was finally enjoying himself—popular, doing well in his classes and on the playing fields. "Besides being good-looking he had one of those one-in-a-hundred smiles," Tim Fornero said.

Mike and Tim roomed together again senior year. Mike was an RA, or resident adviser, in a dormitory called New Building. "The thing about Mike, he was just a no-bullshit guy; he had an honesty about him that was true, and that's unusual in life." Tim struggled with his courses and, like a lot of kids, got bummed out about boarding school life. "We called it the Wooster Blues, and Mike helped you get through that."

For most of senior year Mike dated a white girl from a wealthy New York City family. He acquired his taste in clothes. "He was a real sharp dresser, coming from Boston, I guess," Vince said. "He always had his hair groomed and cut."

They were all close friends, seniors riding high. "We knew everyone in school," said Vince. "Kind of like big guys on campus." There were parties after-hours, and "Mike would be a midnight rambler."

Although outgoing with peers, Mike kept up his reserve in the face of authority. Mike and Tim had an English teacher who wanted his students to build their vocabulary skills. In class he frequently asked for the definitions of words. "Mike knew 90 percent of them," said Tim, "but he wouldn't raise his hand. He'd cover his mouth with his hand and he'd mumble the answer to me or whoever was next to him."

At graduation, Mike's thirty-nine classmates voted him "class flirt," while Vince was named "class jock" and cited for "best legs." Tim was honored as "laziest" and "least organized." Mike was not happy about losing "best dressed" to another classmate who was always wearing clothes borrowed from other kids. Mike thought it was unfair, a form of cheating. "Mike took pride in his clothes," said Tim.

The yearbook staff noted that Mike's trademark was his greeting, the way he said, "Hi." Vince wholeheartedly agreed. "Mike's famous thing was to go up to girls and say, Hi, and they would melt." The staff predicted Mike "will end up as a GQ model." For his yearbook photograph, Mike, dressed in a tuxedo, tilted his head slightly, as he addressed the camera directly, with warm eyes and a groomed look. To Tim, Mike wrote: "You're cooler than I am and you're prettier too. When I am old and gray there's no doubt I'll miss you." To Vince, Mike joked: "It is said that a good friend is an extension of yourself. So stay like me and we'll always be friends." To "Mom and Family," Mike wrote: "I caused a lot of headaches; I made a lot of mistakes but all I really wanted was to make you all happy. I hope this helps. I love you all."

In his essay for college, Mike wrote about his time at Wooster and how he'd come around after an uneven start to finish on a high note of self-respect and confidence. He applied to Providence College, Boston University, the University of Bridgeport, and other schools. He had thoughts about playing college basketball.

The fall of 1984, Mike began at Providence College. Right off, he mostly hung out at the gym and played with the guys on basketball scholarships recruited by the Big East team. Mike did not have a scholarship, but his sophomore year went out for the team anyway. He was a "walk-on" trying to impress the new coach, Rick Pitino. But he did not make the cut. To Mike, a key reason was his grade point average. It was low, and Mike knew the coaches were usually looking to bolster the team's overall GPA by adding a few bench players whose grades were high.

Mike actually found Providence less demanding academically than boarding school, but he did not do well in his classes. "I was distracted," he said. He was enjoying his freedom—sleeping late, playing pickup basketball, and socializing. He worked as a bank teller part-time to earn spending money. "I wanted to have fun," he said.

The biggest thing in his life during college came after his junior year. Mike decided he needed a change of scenery, and, in the fall of 1987, he headed south to Georgia. He enrolled for a semester at Morehouse College, the all-male black college in Atlanta. "I had heard a lot about the school and the area and I just wanted to go there."

While at Morehouse, Mike met Kimberly Ann Nabauns. Kimberly, a year younger than Mike, was from New Orleans. She was a pre-med student at Spelman College, the all-female black college in Atlanta. They fell in love and soon began talking about marriage and family. When the semester ended, Mike returned north, but without firm direction. He was committed to Kimberly, but little else. She was bound for medical school, but Mike wasn't sure what he wanted to do. Sometimes his thoughts returned to his childhood interest in police work, remembering Will Saunders and other officers he'd known as a boy. His sister Lillian took him up on the idea and lobbied him to take the civil service exam.

It was in May 1988, while Mike and Kimberly began planning their life together and Mike was beginning to think seriously about police work, that a man named John L. Smith Jr. sat in his car smoking crack cocaine. He was parked near Fenway Park, home to the Boston Red Sox. It was shortly after sunrise when he began driving away erratically. Two Boston police officers in a cruiser picked up his scent. When Smith drove the 1978 Cadillac through a red light, the police wanted him to pull over. But Smith took off. Soon eight police cruisers raced after him. The chase ended when Smith's car hit a curb, a tire went flat, and the engine died. Officers jumped from their cruisers and surrounded the Cadillac. They ordered Smith to get out. Smith flopped across the front seat. Two officers shattered the front windows with their flashlights. One dragged Smith out and threw him to the ground. Smith was unarmed and limp. Three officers piled on top of Smith while others stood by and watched as he got his licking.

In the beating's aftermath, the thirteen officers stonewalled investigators: No cop saw anything or could explain Smith's injuries. The case of police brutality and the cover-up, which became known as the Brighton 13 case, for the number of officers involved and the station where most were assigned, would haunt the department Mike Cox was planning to join for years to come.

Mike Cox and Craig Jones, in the Tango K–8 car, returned to the club Cortee's shortly after 1 A.M. Craig drove. They headed first to the end of a short street running right behind and below the club, "trying to get a feel, you know, if there were a lot of people up there," Craig said. Nothing had changed since their walk-through earlier in the evening: Cars filled the parking lot, and the club scene was peaking.

The several hundred patrons inside were unaware they were being surrounded by teams of officers in street clothes from the Boston Police Department's anti-gang unit. In the crowd was a young Roxbury man named Lyle Jackson and two of his friends, who were in a card game toward the back of the club.

Outside, Mike and Craig left their original position for a second one. They circled around and drove their unmarked cruiser down Washington Street past the club and then took the first left onto Bowdoin Street. The street went up a hill overlooking the club. In the winter, with trees barren of leaves, "you could see good down through the yards," Mike said. Craig pulled into one of the driveways. He climbed out to see if there was an even better spot to view the club, but there wasn't. Mike got out and took a pee.

They were in radio contact with others in the gang unit—two-man teams, such as partners Joe Teahan and Gary Ryan, which had staked out the club from various vantage points up and down Washington Street. One gang unit colleague named Donald Caisey sat in a cab on Washington Street directly across from the club's entrance. The cab was a decoy vehicle the unit used for nights like this one. Caisey's car had the best view.

Everyone was in place. A few minutes later Caisey radioed that two girls were in front of the club shoving each other. Mike and Craig were on their way in an instant. They didn't want Caisey to out the surveillance cab for this. On the way, another unit radioed that one girl put something down her front, but they couldn't see what it was.

Mike and Craig turned on their police lights and pulled up next to the girls. "What did you put down your shirt?" Craig demanded as he opened his window. The girl hesitated. "You just put something down your shirt—just give it to us." Craig threatened to take her down to the station. The girl reached down her front and pulled out a butcher-sized knife. "It was like a hatchet," Craig said. The girl handed the knife to Craig.

Even though the girl possessed a dangerous weapon, the gang unit wasn't interested in a couple of girls getting into it. They were targeting the street gangs. Craig ordered them to scram. "Go home." The girls, surly, went off into the night in separate directions.

Craig turned around to return quickly to their surveillance spot on Bowdoin Street. They had to be patient, and so they sat in their cruiser and "Just watched—watching and waiting, until two o'clock, until the place closed."

CHAPTER 2

Robert "Smut" Brown

Robert Brown III was his given name, but on the street he was known simply as Smut. It was just after midnight on January 25, 1995, when Smut pulled his maroon Volkswagen Fox onto a side street around the corner from the Cortee's.

Smut climbed out of his car knowing Indira was waiting for him back at their apartment with a shrimp dinner she'd made for the two of them. But waiting for Smut was something Indira had gotten used to a long time ago, back to the fifth grade when they'd first met. Smut was twenty-three now, and Indira's twenty-third birthday was coming up in a few weeks. They had two kids—their first, a girl, was already six.

Smut was okay with Indira waiting because this was Tiny's day—Tiny's birthday. Smut and Tiny had been partying on and off all afternoon, and Smut had agreed to catch up with him at the Cortee's. Tiny was John Evans, and the nickname didn't really fit. He wasn't so tall, an inch or so taller than Smut's five-seven, but he was bull-necked and barrel-chested and topped two hundred pounds. Tiny's hair was shoulder-length but he kept it braided. The two had become friends the summer before. They'd both grown up in Roxbury and Mattapan. They shared an interest in drug dealing, and were now associated with a street gang known as KOZ, an abbreviation for kilos and ounces. Tiny also had a terrible stutter, and Smut felt bad, even sorry, for him, because the stutter was frustrating to Tiny and sometimes made him seem stupid, and Smut knew that wasn't true.

Smut strode toward the front of the club. By his side walked Boogie-Down—or Ron Tinsley. Smut had given him a ride. Boogie-Down looked menacing—had this coldness about him when he raised his eyebrows and stared. He wore a gold-colored ball earring. Like Smut, Boogie-Down was twenty-three, and he had a criminal record for possessing drugs and firearms. He'd violated his probation and was lately trying to lie low by staying at his girlfriend's apartment. He was also packing—a black 9mm, semiautomatic Heckler & Koch pistol. Days before, he'd gotten into a beef with a few guys in his girlfriend's building, and so he was carrying for protection. With bouncers stationed at the club's entrance, though, Boogie-Down left the gun in Smut's car.

Smut was familiar with the Cortee's—just as he knew the area—but he didn't much hang out there. He preferred the Rose Club or Conway's in Mattapan. In fact, of all of Boston's neighborhoods, Smut was most comfortable—and felt most safe—in Mattapan, the southernmost neighborhood before crossing into the suburban town of Milton. Mattapan was where his mother lived, and where Smut used to live with her.

The Cortee's was busy. People milled at the front door. Smut crossed the street. He was dressed in brown jeans made by Guess?, a gray top, and a bulky, brown leather jacket. He wore a gold-colored watch and a gold-colored necklace with a square plate. In his pockets he kept his car keys on a BMW chain, his cell phone, and $795 in cash.

Smut was feeling good, or "nice," as he liked to say. He'd been sipping E & J's "Cask & Cream" during his day spent riding around with Tiny Evans. Smut liked the sweet taste of the creamy liqueur with its hint of butterscotch. He knew who'd be inside—plenty of girls, plenty of other dealers, plenty of guys from rival groups, or, as the police liked to label them, gangs named after a city street:

like Castlegate, Humboldt, and so on. Hopefully, Tiny was already there.

Smut flashed an ID at the door and went in.

By definition, "smut" is a noun with several meanings—a particle of dirt; or a smudge made by soot, smoke, or dirt; or an obscenity. Negative connotations aside, the nickname was actually a term of endearment. One of his mother's girlfriends had come up with it. He was five or six at the time, the kind of boy who could not sit still. He had a knack for turning the family's apartment upside down. One holiday season he crawled under the Christmas tree and toppled it. Another time he tried to make breakfast for his mother, Mattie, but didn't have any idea how to do it. He presented her with a concoction of peanut butter mixed with milk and anything else he could find in the kitchen, leaving a huge mess for her to clean up. But it was hard for Mattie to stay upset with her son, and it was during one of the boy's well-meaning messes that her girlfriend laughed, looked at him, and exclaimed, You just Smut!

It stuck—a nickname born from soul and the latest mess he'd made. When spoken, young Robert heard only warmth, and he embraced his new name. "I guess I was like Dennis the Menace," he said. "He didn't mean no harm. He had a good heart. He was just always getting in trouble." His mother would always still call him Robert while his father tended to call him Bob. But to everyone else in the projects he was Smut.

Mattie and Robert Brown Jr. were living in the Franklin Hill housing project in Dorchester when Smut was born on June 26, 1971, although he was not their firstborn. Living in Georgia, near where Robert was from originally, the couple had a boy named Bobby in September 1965. He fell ill with pneumonia and died in the hospital three months later. Mattie was disconsolate. The couple tried again and had twin girls, and then a third daughter was born in 1968. Smut was born after the Browns moved to Boston. Mattie decided to name him Robert after the first Bobby she'd lost, and Smut grew up a "mama's boy." He was the first to admit it. "Mama really took to me."

When the Browns moved into the Franklin Hill housing project it was nearly two decades old, a complex of nine three-story build-ings built on the rocky terrain and ledges just south of the city's sprawling Franklin Park. The red-brick buildings were clustered around concrete courtyards, asphalt parking lots, and patches of grass. The Browns' unit at 11 Franklin Hill Avenue was located in the corner of one of the courtyards. The entry door, painted gray and made of heavy metal, led to a set of stairs. The Browns' apart-ment was on the second floor, one of four off a windowless landing. The subsidized rent was $40 a month.

From his top bunk, Smut could look out into the inner court-yard and see anyone approaching their entry. He saw lots of fist-fights. "Every day there was a fistfight," he said, or so it seemed to a little kid. One time he was in the entry with one of his sisters when a man from the project began to bother her. The man would not let his sister go past him. She pushed, and they struggled. Smut was scared. When he was older he wondered if the man had been trying to rape her. Beyond a flagpole on the courtyard's far side was a play-ground where Smut's sisters took him when he was a toddler. The Dumpsters they passed made a lasting impression. "Rats flew out of them in the summer."

Smut daydreamed about growing up and becoming an astro-naut, and, for a time, he was into what he later referred to as the "firefighter thing." "Robert was a very active kid," his mother said. "He used to jump in the air and do this flip, and scare me so bad." Smut had lots of cousins and "cousins," the kids of his parents' friends. Families were always getting together. "Mama Janet" Jackson, for example, lived with her family on the other side of Franklin Park near Humboldt Avenue. Smut's aunt would take a carload of kids over to play at Mama Janet's, or Mama Janet would visit the project with her sons. The boys played hide and seek, tag, or football. Smut and Mama Janet's oldest son, Dino, were the same age, while a younger son, Danny, was almost two years younger. Danny was a nickname; his given name was Lyle—Lyle Jackson.

Smut's father was a model of steady work. When the family first moved into the project, he worked as a packer in the shipping department of the Vanity Dress Company in downtown Boston. Laid off, he quickly went to work for a florist and then, just before Smut was born in 1971, he got a job at Doherty, Blacker and Shepard driving a lumber delivery truck. "He worked there forever," said Smut. Bobby Brown worked hard and was proud of his long service with the company. He had a photograph taken in late 1976 of his flatbed DBS truck parked in the company yard, loaded with lumber. The black-and-white picture of the truck—just a truck, no person in sight—was kept in a large envelope filled with photographs of birthdays, picnics, and other family moments.

Unlike other kids Smut knew, his father was home, the head of the family. But Smut and his father were not close, and early on Smut got the idea his father did not like him. Smut felt his father was distant and hard on him. "He was there, but not there," Smut said. "He was always yelling at me." Smut would complain to his mother, saying his father didn't love him. Why you sayin' that? Mattie would ask. She tried to reassure her son, but she also knew he had reason to feel the way he did. "I think Bobby loved Robert," she said, "but had a poor way of showing it." Her husband was abusive. "Robert was afraid of his father." The relationship only worsened as time went on.

Bobby Brown did want more for his family. He began looking at houses, and by the late 1970s, he and Mattie found one a few miles south of the housing project in Mattapan. The seller of 231 West Selden Street was the federal government—the department of Housing and Urban Development, or HUD—which had taken control of the home in 1977 and was selling it off as part of a national housing program. The price was $24,000. Under the program, if Bobby and Mattie Brown came up with $750 in cash, they could finance the balance. The couple seized the chance. They took out a thirty-year mortgage for $23,250, with monthly payments of $195.53.

West Selden Street was long and wide. The top of the street, right off the busy and commercial Morton Street, had mostly two-family homes tightly packed on small lots. Farther south the street opened up—larger lots and more single-family homes. Behind the homes on the even side of the street were the wooded grounds of the Boston Sanatorium, which the city developed in the early 1900s to care for the poor suffering from tuberculosis. Bordering the sanatorium's fifty-one acres gave this end of West Selden Street an almost suburban feel to it.

Two thirty-one West Selden Street seemed like a three-story home, but that was because it was built into a rocky slope, meaning the basement was above-ground and at street level. The front door was up a flight of fourteen steps. There was a tiny yard on the right side, atop the ledge, in the shade of a half-dozen oak trees. The house itself, and the yard, were a mess. The Browns went to work. They installed yellow aluminum siding and cleaned up the yard. "I made it nice, so the kids could play there," said Mattie. Her husband put up a picket fence around the yard and installed outdoor lighting.

Leaving the project was a big deal. "I had my own backyard now," said Smut, who was eight years old when they moved. Instead of the smell of urine, he said, "the air smelled clean." Smut also noticed, "The roaches were gone." Smut sometimes rode a city bus up Blue Hill Avenue to visit his friends from the project, but he also took to the woods of the nearby sanatorium, riding his bike on its paths with his new friends.

Mattie tried to be there for her son. Having gone to work driving a city school bus, she parked the bus at home and used it to take Smut and his neighborhood friends on outings. In the summer, it might be a day trip for a swim at Hogan's Pond in Milton, or to ride the roller coaster at Riverside Park in Springfield, Massachusetts. She'd take Smut and his friends on the bus to Providence to attend a concert.

Smut was in the fifth grade riding the bus to elementary school one day when he saw a girl he could not take his eyes off. "It was her hair, man, her silky hair. Everything about her, she was so smart." The girl was named Indira Pierce, and she was also in the fifth grade. Slowly, on bus rides, the two became friendly, and then Smut made his move in the sixth grade. He passed Indira a note on the bus. It had two questions on it:

"Do you like me? Circle Y or N.

"Do you want to be my girlfriend? Circle Y or N."

Indira circled yes to both. "It was on!" said Smut. They were together from that moment on—all through school and into adulthood. Indira was an anchor in Smut's life.

They rode the same yellow bus from Boston, but actually went to different elementary schools. Both were enrolled in the state-funded METCO program, where kids from Boston were bused to schools in the suburbs. Mattie Brown wanted Smut to have a chance at a better education than the troubled Boston school system would provide. "The suburban schools had more to offer," she said. "Kids were more advantaged out there."

Smut and Indira went to school in Wellesley, Massachusetts. Wellesley was known as a W town, the nickname for a trio of affluent, mostly white communities just outside the city: Wellesley, Wayland, and Weston. The daily bus ride from Smut's house to school was just about eighteen miles, but in so many ways Mattapan and Wellesley were different planets. Once, a teacher took Smut aside and talked to him about his nickname. She said Smut was demeaning and not a good nickname, in terms of his self-esteem and identity. She encouraged him to drop it.

Smut listened respectfully. "When I got home from school I said to everyone, 'I don't want you callin' me Smut anymore. No more! My teacher said Smut is a dirty name.'"

Mattie was taken aback by how upset her son was. She told him if that's what he wanted, so be it. But then her girlfriend, the one who'd first come up with the name, would have none of it. "She starts sayin', 'We call you Smut, you like it or not! You is Smut.'"

Smut later came to see his teacher was well-intentioned but off the mark. He saw it as an example of a culture clash, the teacher's misunderstanding of street talk and black vernacular, or Ebonics. "Like 'phat,'" said Smut. "'Phat' doesn't mean fat, ugly. 'Phat' means cool. And 'Smut' doesn't mean dirty. 'Smut' means love."

In Wellesley, Smut became friends with a boy his age named Derek Roman. In the eighth grade, the two got jobs at a local supermarket as "bag boys." For the skinny kid from Mattapan, the job meant status. "That was the biggest thing," said Smut. "I was the only kid in eighth grade with a job." Indira was impressed. "She was really on me then," he said. "'My baby got a job!' she'd say."

In junior high Smut occasionally spent the night at the Romans. He sometimes tagged along on family outings, like fishing, and envied the boys' closeness with their father. "That's what I wanted with my father."

At home, Bobby Brown made clear his disappointment with his son and would hit him regularly. One time his father yelled at him for putting too much milk on his cereal. Another time he got whacked when his father found fault with the way he raked leaves. "He'd snap," said Smut. "He'd tell me, 'You ain't ever gonna be shit.'" Mattie tried to referee and shield Smut from her husband's habit of demeaning him, but to no avail.

It was true Smut did not shine in the classroom. He was much more likely to have his head in a comic book than a schoolbook. He never earned good grades. Mattie had her son tested for learning disabilities, but none was diagnosed. "He was just so itchy—he couldn't sit still," she said. Smut was disruptive. "He was the class clown—always doing things to keep the other kids laughing." The behavior led to umpteen meetings at school. "I always heard the same thing," Mattie said. "Robert was a clown and troublemaker, but they liked him." Teachers told Mattie that Smut listened to them when scolded and told to stop the horseplay. "He wouldn't give his teachers any lip," Mattie said, "but when they turned their back, he was at it again, always fooling around."

Clowning around in class was one thing. On the street, the stakes were higher. Mattie tried to steer Smut straight. "I was always talking to him and he was listening and respectful and then behind my back he'd do something different with his friends."

When Smut was fourteen he was arrested for the first time—charged with stealing a car in March 1986. "I had a habit of hanging out with older kids, eighteen and nineteen." He watched them steal cars left running by owners warming them up for work. But when Smut tried copying them, he got caught. His mother came to the rescue; she helped convince the juvenile court to give her son a break; the case was dropped.

Smut then had a few other run-ins in "juvie," including one involving Indira's mother. Indira was a repeat runaway from home, staying with Smut in the Browns' unfinished basement bedroom. "She'd come looking for Indira," Smut said. Trying to keep them apart, she complained to police that Smut was contributing to the delinquency of Indira, a minor. "She didn't approve of our relationship," Smut said. But that case also was eventually dropped. His chief defender—always—was his mother.

Smut straddled two worlds—kidhood and adulthood. The little kid in him was crazy about his comic book collection. "I had like sixty thousand comics." His favorites were Marvel Comics' Spider-Man and the team of superheroes known as the X-Men. He wrapped covers in plastic and hung them on his basement bedroom wall. Then there was the first taste of drugs and drinking. His early steps on the wild side included the missteps that come with being a novice. The first time he inhaled marijuana when he was fifteen, he gagged. "I thought I was going to die." He was alone in his basement room when he lit the joint. He choked and completely freaked out. He ran upstairs. "He was all paranoid, all hot and spitting," Mattie said. Smut wanted his mother to drive him to the hospital. "Robert was begging me." Mattie refused. "I told him to call an ambulance. I said, 'I told you not to smoke.'" In a panic, Smut dialed 911 but hung up before the dispatcher understood the nature of the emergency. Instead of an ambulance, a police cruiser pulled up in front of the house. "They were asking, 'What's wrong? What's wrong?'" said Mattie. "I told them Robert smoked some weed and he's hallucinating, like he's gonna die." The officers studied Smut. More amused than anything, they told Smut to take a shower.

The clumsy start aside, Smut was soon getting high regularly, inhaling without a hitch. He complemented the weed with chocolate-flavored liqueurs and wine coolers, while never taking to hard liquor. He was no longer going to school and hardly saw the Wellesley family he'd met through the METCO program. He was barely seventeen, but he fashioned himself as an up-and-coming small businessman with an entrepreneurial streak. When he tried cocaine, for example, he didn't much like it, but he did like the drug's earning potential. "So I started selling," he said. To get started, he and a friend each chipped in $75 to buy an "8-ball" from a supplier. They cut the coke on a plate into thirty chips, or "jumbos." They wrapped the jumbos in tinfoil and sold them for $10 apiece—doubling their investment when they were done. The two set up in a crack house about a ten-minute walk from Smut's house on West Selden Street.

When he suspected his partner was keeping more than his share, Smut decided to split. He began dealing by himself on the street. He picked a spot along Blue Hill Avenue near a skating rink, right across from the neighborhood police station known as B-3. He quickly learned he had to pay to play—meaning pay off a police officer who was notorious for hitting on the street dealers working in Mattapan and Roxbury. Smut saw the fee as a business expense in the stream of illegal commerce that made him good money. "Got me better clothes, better sneakers," he said. He also discovered he had talent for sales. Being lyrical of mind, he'd come up with a winning ditty for the pitch he made quietly on Blue Hill Avenue: "If you pass me by you won't get high."

Nineteen eighty-eight was a big year for Smut. For one, he and Indira were past puppy love. They were inseparable and acting beyond their years of seventeen and sixteen. By summer's end Indira was seven months' pregnant. Fatherhood was on Smut's mind. But it wasn't his only concern. The year was also defined by deepening troubles at home and in the street.

Smut's parents smoldered in anger and tension. Mattie, saying she could not take her husband Robert's abuse anymore, moved out. "I'd just had enough," she said. Her husband, she said, was a "control freak," always yelling whenever she went out with friends—Where the hell you goin'? They argued about his drinking. "I wanted him to stop," she said. Robert never hit her, she said, but when he drank he was "abusive with his mouth. I got sick of hearing it." Mattie moved in with another man.

Mattie was Smut's anchor, and suddenly she was gone. He took it personally. "This hurt me a lot because I didn't understand why." Mattie was no longer there to stand between Smut and his father. "My father changed for the worse and my home life became a living hell." Smut got mouthfuls from him—how he was no good—and Smut wanted out of his father's line of fire. He stayed at friends' apartments. He sold coke. He spent his time with Indira.

"I was running the streets." There were also secrets and lies. Smut and his older sisters knew where their mother was staying, but their father did not. They lied as he angrily tried to figure out Mattie's whereabouts. To see Smut, Mattie picked him up at one end of West Selden Street in her new friend's car. She'd take him out to dinner. Smut pleaded with her to come home. Mattie assured her son that she loved him. Eventually, six months after she'd left, Mattie returned home. "I came back to my children."

Nineteen eighty-eight was also the year Smut's luck on the street ran out. Mattie had been home only a little while when one of Smut's friends came by the house late one night. It was October 5. Smut climbed into the friend's car and the two drove to Canton, a suburban town located south of Boston. They left their car and, in the dark, snuck to the back of the Coleman's Sporting Goods store. They broke a window, got inside the store, and grabbed some guns. Driving away, north on Route 138 toward Boston, they were pulled over by the local police . They were arrested the moment police saw the automatic weapons.

Smut had broken into the big time. Within weeks, he was indicted for breaking and entering with the intent to commit a felony, malicious destruction of property, two counts of theft of a firearm, and possession of a firearm without a permit. The charges carried heavy prison time. "Before this, when he got in trouble, he got out of trouble. I posted bail, whatever," said Mattie. "But I told him there's going to be a time when I can't help." Smut's timing also could not have been worse. On October 12, seven days after his arrest, Indira gave birth to their daughter. They named her Shanae.

Smut found himself juggling court appearances with hospital appearances. "I was a knucklehead, the things I did," said Smut. But self-awareness did not stop him. Over the course of the next twelve months while the gun case was pending, Smut ran up a slew of new criminal charges. In December, he was charged with receiving stolen goods. Four months later, on March 14, 1989, he was caught in the suburban town of Norwood popping the ignition on a car. The next week, he and three friends were arrested by police in Waltham, Massachusetts, driving two stolen cars, a Chevy Camaro and a blue Oldsmobile. The next month, he was busted by Boston police for coke possession. In August, he was caught trying to use a screwdriver to steal a car in the West Roxbury neighborhood of Boston, and later that month, he was stopped by police for driving while his license was suspended. In October, he was arrested again for coke possession.

Smut faced seven new court cases in the year since his arrest for breaking into Coleman's. Mattie Brown scrambled to find lawyers and keep her son out of jail on bail. Then, in the midst of his year of living criminally, Smut made his biggest mess ever at home. He burned down the house—literally. Smut and Indira were living in his basement bedroom. She had run away from home after Shanae was born. The baby slept in a bassinet in Smut's parents' bedroom, while Smut and Indira occupied the partially finished bedroom. The room was unheated. Smut's father had warned against using a space heater, but Smut ignored him. The heater cut the nighttime chill of late winter.

Smut wasn't certain how the fire began. "I must have thrown a pillow on the space heater." He awakened to Indira's screams, Fire! Fire! His first thought was how his father was going to kill him. Smut tried throwing water on the flames, but it didn't help. The flames spread, running quickly up the walls, fueled by the comic books he'd hung like wallpaper. Smut ran upstairs.

Mattie, a light sleeper, heard her son outside her bedroom door. "He came into the bedroom and then walked back out into the hall, pacing." Smut was afraid to tell his father what had happened until Mattie yelled, "What's the matter, Robert?" Smut blurted out about the fire; his parents leaped from the bed. Smut grabbed Shanae. Everyone got out safely. But by the time firefighters extinguished the blaze, the house was uninhabitable. The smoke damage was extensive. They had to move out while the house was repaired. Smut's family moved in with his aunt's family in Hyde Park. Indira moved back with Shanae into her mother's.

Smut's troubles were now not the kind his mother, Mattie, could be expected to straighten out.

This was especially true for the pile of criminal charges he'd amassed while on bail in the Coleman's burglary. There was no longer a way out. One year after the break-in, Smut stood in court on October 17, 1989, and admitted his guilt. The judge sentenced him to serve two and a half years in the House of Corrections. If there was any good news to pleading guilty, it was the resolution of the other seven criminal cases. That's the way the system worked—once Smut pleaded guilty in the big case, the other cases were eventually disposed of with little additional damage. Some charges were dismissed, while Smut pleaded guilty to others. The new sentences ran concurrently to the time he was already serving. Still, the cleanup came with a price. To pay for her son's legal bills, Mattie took out a second mortgage on their house in Mattapan in the amount of $50,000.

Smut entered the prison system at age eighteen. He was released in July 1991 after serving twenty-one months of his thirty-month sentence, shortened for good behavior. He was now a twenty-year-old ex-con. But little was changed—in him, in his world. He moved back into the house on West Selden Street, which had been renovated. Indira rejoined him in the basement bedroom, and their second child was born on March 13, 1992. They named the boy Robert Brown IV, and soon he was nicknamed "Little Smizz."

Smut resumed the livelihood he knew best—dealing coke. He spent his days getting stoned and dealing the drug on the streets of Mattapan, Dorchester, and Roxbury, though he had a knack for getting out of the trouble that inevitably came his way. He was soon arrested by Boston police, but was later found not guilty of the drug-dealing charge in Dorchester District Court. In early January 1992, he was arrested again, but he beat that drug-dealing charge too, winning another ruling of not guilty.

Hosting Hip-Hop Night was a display of business acumen by the Cortee's. Generally speaking, Boston was not a destination for rappers and hip-hop shows. The shows that did make it to New England took the stage at the Centrum in Worcester or the Providence Civic Center in Rhode Island.

Walking inside, Smut and Boogie-Down were swallowed by the club's darkness. The dance floor in the center of the room was full. The bar along the right side was deep with patrons. The few tables were all taken. The DJ in a booth straight across the room was playing everything—from Notorious B.I.G., the king of hip-hop, to Wu-Tang Clan to Nas, the street poet. Lots of "gangsta rap," vicious and raw, violent and drug-fueled.

Smut spotted Tiny Evans.

Tiny was with Marquis—or Jimmy—Evans, Tiny's younger brother. He was the biggest of them all—more than six feet tall and weighing 220 pounds. Smut hardly knew Marquis, who was his age, twenty-three. And Marquis had just gotten out of prison—convicted at age seventeen of using a sawed-off shotgun in an assault case. The one thing Smut knew was Marquis could be a hothead, which Tiny sometimes manipulated to his advantage.

Tiny saw Smut and hurried over. Tiny had spotted a kid named Little Greg who was affiliated with the Castlegate Street gang. "Tiny was saying, 'Little Greg is in the club, Little Greg is in the club,'" Smut said. "He was talking a mile a minute."

Tiny and Little Greg had a beef going back a couple of years—beginning when Tiny ripped Little Greg's chain right off his neck and kept it. Then the previous summer Little Greg got some revenge. Tiny told Smut he was getting his hair cut when Little Greg burst into the barbershop and fired a shot. The next time Smut saw Tiny he was walking with a cane. "He got hit near his scrotum." Not surprisingly, neither event was reported to police. They were matters for the street. Now inside the Cortee's, Tiny and Little Greg exchanged looks. Smut saw that Tiny was monitoring Little Greg's whereabouts. Smut reminded Tiny it was his birthday. "Leave it alone," Smut said.

Looking around, Smut observed friend and enemy alike. But among the foursome—Smut, Tiny, Marquis, and Boogie-Down—he felt secure. The group stood at the bar. Boogie-Down spotted his girlfriend and snuggled with her. Marquis was broke but wanted to buy a round of drinks in honor of his brother's birthday. He had the gall to ask Tiny to loan him $20. Tiny couldn't believe it, but dug into the pocket of his blue jeans, where he had a roll of more than $700 in cash. Drinks were on Marquis.

The songs worked loud and hard on Smut. He ordered a drink, another smooth Cask & Cream. Smut loved rap. He saw himself as a budding lyricist and eventually would go from toying with words and beats inside his head to writing them down on paper.

They were mostly autobiographical lyrics like:

> *I had a Daddy who was crazy so I lost my patience*
> *That's when I hit the street, searchin', hurtin', wantin' salvation.*
> *My occupation was me cuttin', puttin' rocks in a bag. . . .*

It was a verse from a song "Our Hoods" by Smut Brown, whose hook went:

>*Ya'll don't know what it is*
>*To grow up in our hood (our Hoods!)*
>*Ya'll don't know what it is*
>*To see the things that we would.*

CHAPTER 3

Kenny Conley

When Kenny Conley arrived that night at the station in the South End of Boston to work the overnight shift—known as the "last half," from 11:45 P.M. to 7:30 A.M.—he first went to his locker on the second floor to get his equipment squared away. Then he walked back downstairs to read some reports and talk to the guys coming off duty to see what kind of night it had been. That's when he learned he was without his regular partner, Danny McDonald, who was out on an injury and not available for duty.

Kenny wasn't all that surprised. McDonald had injured his knee the night before while the two worked an anti-crime unit— the Delta K–1 car—patrolling the district in an unmarked cruiser looking for trouble: drug dealing, prostitution, crimes in progress. "You're out there hunting," Kenny said. The anti-crime cars were considered more pro-active than the "service units" that were directed by a dispatcher to respond to calls for police assistance, ranging from a disabled vehicle to more serious crimes.

The anti-crime units were also different from the police department's elite units that also worked in street clothes. Kenny patrolled only in his district, known as Area D–4, which covered parts of the South End, Back Bay, and Fenway neighborhoods. Officers assigned to either the drug unit or the Anti–Gang Violence

Unit—such as Mike Cox and Craig Jones—had citywide jurisdiction and were free to roam.

The night before, Kenny and McDonald had driven slowly down one of the narrow alleys running behind the townhouses and red-brick buildings that made up the Back Bay, the historic neighborhood that was home to a mix of students, young professionals, and the well-heeled. Kenny was driving when they noticed a car ahead of them, occupied and idling. They watched as two men approached the car, textbook "suspicious activity" for that hour of the night. The officers ran the car's plate. When it came back as a stolen vehicle, McDonald opened the cruiser's door, climbed out, and began walking to the car. That's when the car lurched forward. Instantly, Kenny hit the gas pedal and looped around to cut the car off. Other police units responded in time to catch the two men who fled on foot. The suspects were taken into custody, while McDonald was taken for treatment. He'd been hit in the knee by the lurching car and would eventually undergo surgery to repair the ligament damage.

Kenny learned during roll call that McDonald would be out. His supervisor asked whom he'd like to work with that night in the Delta K–1 car. Kenny looked around the guardroom full of officers ready to go on duty. He spotted Bobby Dwan. He'd never worked with Dwan before, but they were friendly. Bobby had come onto the force in 1990, a year before Kenny did. He was a second lieutenant in the National Guard who had served for six months in the Gulf War in 1991 as a platoon leader in the military police. Like Kenny, Bobby was from Boston, although Bobby grew up in Mattapan on the opposite side of the city from Kenny's South Boston. Bobby was a jock; he was a three-sport varsity athlete in high school—football, hockey, and baseball—and played center for the first line on the police department's hockey team. He was married, with a baby girl and another due any day now, and he lived just outside the city. There was no pretense about Bobby—nothing fancy and no bull—and Kenny liked that.

How about Bobby Dwan? Kenny told the supervisor.

It was done.

Bobby had been scheduled to work a one-person service unit, so he had to run to his locker and change back into the clothes he'd worn to work—blue jeans, sneakers, and the L. L. Bean barn jacket with the green corduroy collar his wife had bought as a gift. He joined Kenny, who already was set to go—dressed in jeans, sneakers, a black turtleneck, an off-white Carhartt jacket, and a corduroy baseball hat with a shamrock on it.

They headed out to the Delta K–1 anti-crime car. Side by side, they were an odd couple: Kenny towered at six-four and weighed 215 pounds, while Bobby was five-three and barely topped 150 pounds. In the city, the big news at the time was the nationwide manhunt for Boston's most famous gangster, James J. "Whitey" Bulger. Under investigation for years, the aging crime boss from Southie had hit the road at the beginning of the month after a corrupt FBI agent tipped him off to a pending federal indictment. Whitey disappeared with a girlfriend, and soon enough the sixty-six-year-old killer made the FBI's Ten Most Wanted List alongside Osama bin Laden.

Whitey was the talk of the town, especially in Southie. But for twenty-six-year-old Kenny Conley, another son of Southie, all the Whitey talk was background noise to personal anguish. His mother had died—on Thanksgiving Day. She'd fallen ill suddenly in October, was hospitalized, and fell into a coma. She never recovered. She was fifty-two.

"I took that hard," Kenny said. He and his mother had been very close. "It was the toughest thing I ever went through." He thought about his mother every day. But he was not about to talk to Bobby Dwan about her. "When I'm on the job, I focus on the job."

The first-time partners left the D–4 station after midnight. Within minutes, the Delta K–1 unit was heading to East Newton Street in the South End to investigate a report that prostitutes were working a street corner despite the sub-freezing cold.

Growing up in South Boston, Kenneth Michael Conley always wanted to be a cop. His uncle Russ—his father's oldest brother—

was on the force and worked for years at the same station where Kenny was eventually assigned—Area D–4. As a boy, he had been impressed by his uncle's uniform. "I'd see my uncle coming home, in his uniform with his partners, coming to see my father, and it excited me." In addition, Kenny's boyhood perspective on his uncle's duties neatly fit with the Southie virtue of help thy neighbor. "I like to help people," he said. To a question in his eighth-grade yearbook asking what he would be doing twenty years later, Kenny's answer was: "Boston Police Officer."

His modest upbringing was one of the typical Southie stories unfolding within a few blocks of home. When he and his twin sister, Kristine, were born on December 11, 1968, his parents lived in a third-floor walk-up at 599 East Fourth Street with their first-born, Cheryl. His parents, Ken and Maureen, or "Moesie," were both from Southie. They'd met when their respective "crowds" crossed paths. Maureen; her oldest friend, Peg O'Brien; and their other friends hung out at Frank and Rosie's on N and Sixth Street. Kenny and his pals hung out at a spa one block away, on N and Fifth. Maureen and the girls would go to the spa for pizza and to play the jukebox, and the guys in Ken's crowd would follow them back to Frank and Rosie's. "Before long it was one crowd," said Peg O'Brien. Ken, who was four years older than Maureen and a high-school dropout, worked as a truck driver and later as a track worker for the Massachusetts Bay Transportation Authority, or "the T." Maureen worked for an insurance company, but quit when Cheryl was born. Within a couple of years, though, Maureen and Peg got part-time jobs at Gillette, headquartered in Southie, testing new deodorants. They'd sit with a "panel" of other women in hot rooms with pads under their sweating armpits to test the effectiveness of the deodorant. The two friends worked as a tag team, alternating between work and home. While Maureen worked, Peg watched Cheryl; when Peg worked, Maureen watched Peg's daughter.

"Kind of like Lucy and Ethel," Peg O'Brien said. "I think it was for about ninety minutes a day. We got paid about $35 a week."

When the twins Kenny and Kris were born, Maureen decided to stop working again and stay at home with her three kids.

The Conleys lived in a four-bedroom apartment with a single bathroom, one block from East Broadway, which, along with West Broadway, was Southie's main commercial street. The two Broadways ran the length of Southie, from Boston Harbor on the east to a bridge on the west side that connected the neighborhood to the city. For a third-floor apartment, the Conleys' home did not have much of a view. They looked out onto the asphalt parking lot of the telephone company building that occupied the entire block from East Broadway to the side street—H Street. The far side of the parking lot actually rose uphill, an incline leading to the back entrances of some retail businesses on East Broadway. Kenny Conley called the tiny hill Tar Hill. In the winters after a fresh snowfall, he and his pals used it for sledding. The "trail" began atop a sliver of grass, ran under an iron railing, and then across the asphalt lot. The chain-link fence at the sidewalk served as a safety net, stopping their sleds from shooting out onto the street.

The Conley homestead was only 3.8 miles from where Mike Cox and his family were living in Roxbury—but the two neighborhoods were a world apart. Southie was overwhelmingly white and Irish—and had been since after the Civil War when the first wave of Irish immigrants moved into the area.

In the other three-decker apartments and row houses surrounding the Conleys lived families much like their own, where the breadwinners mainly worked in the trades, the public utilities, Gillette, "the T," or the police and fire departments. The median family income when Kenny was a toddler was $11,200 annually, and the majority of grown-ups never went to college. It was blue-collar through and through.

In the beginning, meaning back to the American Revolution, the grassy and hilly peninsula jutting into Boston Harbor was ideal for grazing livestock. In March 1776, the Colonial forces, led by General George Washington, used it as a base from which to drive the British out of Boston. By the early 1800s, South Boston formally became part of Boston, connected to the downtown by the new Dover Street Bridge.

Given its geography—nearby but separate and isolated—Southie became a convenient location to build the city's prisons, hospitals for the mentally ill, and poorhouses. Iron foundries, machine shops, and shipyards on the waterfront all sprang up.

The potato famine from 1845 to 1850 that devastated Ireland triggered a massive exodus to the city, first to the tenements of the North End and then, in the decades following the Civil War, to Southie. The Irish eagerly took jobs in shipbuilding and along the waterfront unloading freight ships. The women traveled across the bridge where they cleaned the homes of the Brahmins on Beacon Hill. Life revolved around work, family, and the many new Catholic churches opening throughout a neighborhood that was only 3.1 square miles. In time, it was said that Southie was a "state of mind."

"For those born and raised there, South Boston was a warm, friendly, comfortable community where people knew one another, shared the same values, enjoyed the same pastimes, and were safe from outside contacts and alien influences," wrote the historian Thomas H. O'Connor, a history professor at Boston College. "Southie pride" became a powerful force embedded in the clannish, tight-knit neighborhood—an us-versus-them mentality between the neighborhood and the rest of the city, or world for that matter.

Kenny Conley quickly came to embody that pride and the neighborhood's cultural emphasis on staying put rather than breaking away. "I did see myself growing old, sitting right down here, in South Boston. I've always said, and I think it's the sentiment of everyone around here, Why leave God's country?" From early on, in addition to working as a police officer, he'd say his dream was to marry and have a family. And his idea of making it was to find a house in the densely built neighborhood that actually had a garage and a driveway. "You know, something I could take a snowblower to."

Kenny was nine years old when his parents realized their own dream and became Southie homeowners. His father paid $10,000 in cash for 78 H Street—the third in a cluster of six row houses they could see around the corner from their apartment. The house had white aluminum siding with black trim. There were four bed-rooms—two on the second floor and two on the third—and two bathrooms, a luxury for a neighborhood where a single bath was the norm. The Conleys closed on the house on November 3, 1977, but could not move in. No one had lived in the house for years, and the inside was a wreck. "We did a total gut job," said Kris, Kenny's twin sister. They made the move three months later, just before the Blizzard of '78, the nor'easter that dumped more than twenty-seven inches of snow on the city between the morning of February 6 and the next night. The renovation was not finished and the interior was always a work-in-progress. This was because Maureen Conley decorated the house herself, and then she'd do it over again.

Kenny's bedroom was on the third floor at the top of the stairs. It had a sliding French door opening into a brown-paneled room his parents had carpeted wall-to-wall with navy-blue carpet. His mother chose a color scheme of red, white, and blue, and every-thing was matching. She decorated the room with ceramics she and her girlfriends made at a local shop.

The second bathroom was on Kenny's floor. It had a rear window opening onto the second-floor roof. His parents stored an eight-foot ladder on it, and during the summer they'd climb out the bathroom window and use the ladder to get to the third-floor roof. They kept folding chairs up there for sunbathing. To the north, there was a view of the sprawling Edison utility plant and the ship-ping terminals along the waterfront. To the south was a steeple from the Catholic church on the next block.

The family room, or den, was in front on the second floor over-looking H Street. This was where the kids hung out, where the large TV console was stationed, where Maureen was at her most imaginative. One wall featured a fake fireplace that, when turned on, made crackling sounds and flickered with phony flames. "The room had a country feel," Kris said. "It sounds tacky and crazy, but it worked."

The move to 78 East Fourth was hardly a big one for Kenny, his two sisters, and their parents—just around the corner from the apartment on East Fourth. They now faced the side of the telephone company building instead of looking out onto Tar Hill, the parking lot behind the building. But there was one key difference—their stretch of H Street was atop one of Southie's hills. Looking south on a clear day, Kenny was able to glimpse the ocean waters of Old Harbor off Carson Beach, the main beach in Southie.

The corner at H and Fourth Streets, along with the next corner—H and Fifth—defined Kenny Conley's universe. "Nothing's changed since I lived here," he once said while standing on H Street as a grown man. As a little boy Kenny made friends who then became friends for life. His best friend, Michael Doyle, lived on East Fourth in a house located between Kenny's old apartment and his new house. "In front of his mother we called him Mike, but 90 percent of the time he was just Doyle," said Kenny.

Kenny, Doyle, the other Mike—Mike Caputo—and other pals turned the corner of H and Fifth into their own Fenway Park for wiffleball. Using the same kid ingenuity that had led to sledding down Tar Hill, they made the four street corners the bases. Home plate was located at the southwest corner—which meant hitters drove the wiffleball slightly uphill and upwind. Games were interrupted by passing cars almost always occupied by a neighbor or relative. Between games they'd take a break and wander into the variety store at the northwest street corner (third base). The store, where Kenny's parents often sent him to buy bread or milk, was owned by Mike Caputo's parents. Kenny usually bought a Pepsi and Reese's peanut butter cup.

The boys owned the corner—a hangout after school and during the summers. In the ninth grade, Kenny scribbled an ode to his friends and their place inside the closet door in his bedroom. It read: "H + Fifth . . . #1." Under that, Kenny then drew a shamrock and wrote "Southie" underneath the shamrock, and then he wrote his friends' names.

Sports were king. Kenny and his friends played wiffleball, baseball, pickup football, just about any game they could come up with. Lots of kids in Southie laced up ice skates and became hockey players at the neighborhood rink, but Kenny never caught the hockey bug. Right away, his favorite game was basketball. He was always tall for his age, an advantage Kenny had right into adulthood, when he topped six feet and kept going.

But his height did not necessarily mean Kenny was the hot player everyone wanted when it came to choosing up sides. "I wasn't usually the first pick," he said. Kenny was not what was called a "skill player." He wasn't a fancy passer or ball handler whose slick moves faked and fooled players on the other team. He didn't possess a sweet shot, either beneath the hoop or from far away. Kenny was the opposite of finesse. "My game?" he once asked rhetorically. "I don't got game." He joked: "I'm not known for anything except for standing there." His game was physical, rugged, and without nuance. He pulled down rebounds. In fact, his game mirrored his personality—straight-ahead and no bull. There was never anything slick about Kenny Conley. On and off the court, what you saw was what you got—a hardworking, unpretentious kid without a shred of guile.

Kenny played most of his basketball one block away from his house in the second-floor gym of the Gate of Heaven Church. The brick church was built in 1863 during a period when the Irish immigrant population was exploding and spreading east across Southie toward City Point. Kenny practically lived in the hall, playing basketball year after year in the church's Catholic Youth Organization, or CYO, league. He was ten years old when a young priest named Father Kevin Toomey came to Gate of Heaven. Father Toomey ran the CYO programs, and he became a mentor to Kenny and his friends who hung out at "Gatie." Father Toomey drove the boys to their away basketball games. For a couple of years when Kenny, Mike Doyle, Brendan Flynn, and Bobby McGarrel were teenagers, they picked up $10 each from the father for "breaking down the hall" after Bingo Night and getting it ready for Saturday CYO basketball. The boys worked late, and Father Toomey often came by to check on them. He would sometimes toss around a football to break up the monotony of folding tables and chairs at midnight. "He kept us straight," Kenny said. When Kenny was a high school senior in 1987, he was awarded the parish's Catholic Youth of the Year Award, and a plaque inscribed with his name was hung in the Gatie gym. The winner the year before was his best friend, Mike Doyle.

Kenny had everything he wanted within a five-minute walk from his house—his friends, school, church, the Gatie gym, the playing field at the corner of H and Fifth Streets, and the Italian cold-cut grinders at Mike Caputo's parents' variety store. His boyhood was simultaneously unexciting and fulfilling. "I just did what I was supposed to do," he said. His horizon expanded a bit when he and his friends got their drivers' licenses. "We'd drive to Castle Island to Sully's," he said, "which has the best hot dogs in the world." It was a comment at once serious and comic. Castle Island in Boston Harbor, just off City Point, was connected to Southie by a causeway. In 1970, when Kenny was two years old, the island and the fort built on it during Colonial times were placed on the National Register of Historic Places. It was only about a mile from Kenny's house. But to a boy on H Street, the five-minute drive there seemed really far away.

It wasn't as if Kenny never left Southie. In the summers, his mom took him and his sisters to the Cape. They'd pile into the station wagon and visit Peg O'Brien at her cottage, nicknamed "Grump's Stump." They often went on weekend and vacation trips with their mother's friends—Peg, Twinkie, Nancy, and Arlene. The kids swam and played while the mothers enjoyed "mothers' medicine," frozen lime juice and vodka.

The Conleys traveled to Disney World in Orlando, Florida, when Kenny and Kris were eight years old, and they drove another time to Niagara Falls, where they splurged and stayed at a Sheraton hotel. During summers they sometimes drove seventy miles north to York Beach, Maine, and stayed at the Sands Motel with its large swimming pool. During school vacations, families assembled at spots like The Elms, a ski resort in Manchester, New Hampshire, or the Brickyard, another skiing area in New Hampshire, where Kenny broke his leg when he was twelve.

The one dark shadow was his father's drinking. "It was never really a problem at home or on vacations," Kris said. "But if my parents argued it was about Dad's drinking and his being out and carrying on." Kenny's father had a rough-and-tumble look about him; he was a heavy smoker with tattoos on his forearm; later, he shaved his head and had an earring in one ear. After working all day driving trucks he would hang out in the bars. "You knew when he was drinking, but he was never doing it around the house," Kris said. Their mother wouldn't let him. Over time, the tensions got the better of the couple. The marriage broke down for good soon after Kenny and Kris graduated from high school. Maureen and Ken never divorced, but they never lived together again. And it was during this troubled time that Maureen started drinking heavily. "I knew it was a problem when I saw her drinking at home," Kris said. She saw it as her mother's "mid-life crisis." "She was always a doer, but now she had no kids to tend to, she was upset about the marriage, she had this freedom and was unhappy."

Maureen had been working for some time as a waitress at the Park Plaza Hotel. She'd gone back to work when the twins were in the fifth grade. Having taken her role as a stay-at-home mother so seriously, she actually asked the eleven-year-old twins Kenny and Kris for their permission. "She explained we would only be home alone for about thirty to forty-five minutes between the time we got home from school and when she got home from work," Kris said. "She was all concerned, but we thought it was great." They'd go wild during the brief but daily stay of parenting. "We'd have these blow-out fights," Kris said. But the shenanigans ceased once they heard their mom pushing open the big front door.

When it came to school, Kenny Conley—along with Mike Cox in Roxbury and Smut Brown in Mattapan—was a child of busing, the court-ordered remedy to desegregate Boston's public school system. None of the three boys was ever directly in the line of fire. Their parents joined the legions of Boston parents who, during the busing era, avoided the tumultuous public schools and sent their kids elsewhere. Mike Cox was sent to St. Mary's School in the neighboring city of Brookline, Smut Brown was enrolled in the METCO program and bused to the affluent Boston suburb of Wellesley, and Kenny Conley attended one of the Catholic parochial schools not far from home.

Kenny considered himself a "Gatie," and the Gate of Heaven School was right around the corner, but he and his sister attended elementary school at St. Peter's. The brick Catholic parish school with the tiny asphalt playground was located on Sixth Street, a "commute" of three blocks from Kenny's house. He attended St. Peter's because Cheryl had gone there and his parents liked it. The school was grades one to eight. Kenny's classmates were the same year after year—another stitch in Southie's tight-knit way of life. "Each grade was about thirty-five kids, and I basically went through with the same kids."

He was a freckle-faced boy of five, with a big smile and a mop of hair, when the buses began rolling in 1974. They carried black students from Roxbury to South Boston High School, and they transported white students from Southie to other city neighborhoods. It turned Southie into a war zone. State police patrolled school corridors, riot police flooded the streets, and police snipers took up positions atop three-deckers to enforce the law against the often violent anti-busing protesters. Many in Southie did not deny the school system was segregated, but they found unacceptable a solution that forced students out of neighborhood schools. But to a national television audience the angry confrontations between blacks and whites made Southie seem a hotbed of intolerance. Some of the ugliest moments showed Southie women shouting,

"Niggers go home" at buses filled with black children trying to get to school.

The clash of politics, law, and educational equality was over Kenny's head. But the high school was only a few blocks from his house, and the protests and street fighting were all around. Kenny, Kris, and their mom were eating dinner at a neighbor's first-floor apartment one evening when the front door flew open and a teenager came running through the house. "No one locked their doors back then," said Kenny, "and this kid came in and ran through the kitchen and out the back door. There were a couple of cops right behind him. It was crazy. We watched and went back to dinner."

During the early years of busing some of Kenny's peers were swept up in the anti-busing fervor and joined the demonstrations. Not Maureen Conley's son. "I was the kid, when they were egging buses, I was always coming home." Fourth Street was a route protesters took to the high school for a demonstration, and Kenny, his sister, and their friends were sometimes hauled off the street by a watchful parent. "I can recall being told to hurry and get inside," Kris said. "But I didn't really know why at the time."

For high school Kenny wanted to follow his pal Mike Doyle, who was a year ahead of him, to the Don Bosco Preparatory High School in Boston. Never a star academically, Kenny went to summer school in 1982 so he could get in. He took courses in English and math. It worked. He began attending the Catholic school in September, catching the number 9 bus each morning at the corner of H Street and East Broadway for the ride through Southie and across the bridge into downtown Boston.

The extra effort may have gotten Kenny into Don Bosco, but starting out he was at best a mediocre student. Freshman year he got mostly low B's and C's. Then during sophomore year Kenny began to click—his grades improved steadily. That year and the next he earned mostly B's and A's. He peaked his senior year, both in class and on the playing fields. He played basketball and varsity football, and his grades were so strong he made the National Honor

Society. "It felt good being able to come home having a 100 on an exam," Kenny said. His perfect grades—100s across the board—in his religion class earned him the Religion Award at graduation in the spring of 1987. He also was named a Golden Bear, one of the school's highest honors, awarded for character and leadership. The previous year's Golden Bear was none other than Mike Doyle.

The awards left Kenny feeling a little dizzy. To be sure, he enjoyed them, but he was not used to the attention and did not consider himself "an awards or medals guy." Glory-seeking was not what made him tick; instead, like his mother, he was a "doer." Kenny Conley saw himself as one of the guys who got the job done without fanfare.

Kenny was coming of age in the long aftermath of busing and shifting sands in his hometown—namely gentrification. Slowly, young professionals were discovering the neighborhood's proximity to downtown, its sea breezes, and its water views. But even as the outsiders arrived, Southie's public image remained largely negative. The tumult of busing in the mid-1970s might have long subsided, but Southie had been scarred deeply.

"Although the crisis over busing was a relatively brief episode in South Boston's 300-year history," the historian Thomas H. O'Connor wrote, "it was an unusually bitter and violent period that stereotyped the neighborhood forever in the minds of people throughout the nation as a place where beer-bellied men and foul-mouthed women made war on defenseless black children." The stereotype was ripe for exploitation and would be used against Southie—the sense of loyalty made into a vice, not a virtue.

Kenny would someday experience this firsthand. But in 1987 he was riding on his own modest-sized version of cloud nine. Following the strong finish at Don Bosco, he spent the summer hanging out with friends, driving a delivery truck, and enjoying himself. He lived at home and had few expenses. His parents' marriage was unraveling, but they had stayed friends. Kenny began playing basketball in a new adult CYO league at Gatie. One of the other teams, called the Evans Club, consisted of the Evans brothers, including

Paul, a high-ranking officer in the police department who was twenty years older than Kenny and eventually became police commissioner during the 1990s.

Kenny also was accepted into Suffolk University in Boston. He registered for classes and lined up financial aid and grants. But when September rolled around, Kenny was a no-show. "I just didn't want to go." He decided he'd had enough of school and was talking to his father and Uncle Russ about the Boston Police Department. With their guidance, he filled out an application. He took a police cadet exam. Then, one day in November 1987, Kenny got the call to be a cadet, the first step in his dream of becoming a full-fledged police officer. Kenny was told to report for duty on December 5, 1987—six days before he turned nineteen. Mike Doyle was also accepted into the cadet program.

Kenny's first assignment was working in the traffic division. He was on the job only two months when tragedy struck the department. Heavily armed members of the Drug Control Unit had quietly made their way up the stairs to an apartment on the third floor of 104 Bellevue Street in Dorchester. It was around 8:30 on the night of February 17, 1988. Using what was known as a no-knock search warrant, the plan was to surprise a cabal of drug dealers known to be working out of the apartment. The cops paused outside the bolted steel door and then began smashing their way inside using a battering ram and a sledgehammer. That's when the whole thing went awry. Shots were fired from inside. One of the officers, Sherman Griffiths—thirty-six years old, married, and the father of two little girls—was hit in the head. His partner, Carlos A. Luna, and other cops hauled him out of the line of fire. They tried desperately to treat the wound and resuscitate the burly, bearded eighteen-year police veteran. He was rushed by ambulance to Boston City Hospital and was pronounced dead a few hours later. The police world mourned.

In the aftermath of Sherman Griffiths's death, Police Commissioner Francis M. Roache called the drug unit officers "highly trained and very professional." But as time went on the tragedy erupted into scandal. When it came to prosecuting the man charged in the cop's death, Detective Luna could not produce the confidential informant cited in paperwork to obtain the search warrant. Luna had written on the warrant application he'd obtained probable cause for the raid because "John" had provided him with first-hand intelligence about the drug den. But it turned out there was no John; he did not exist.

The drug unit's unlawful practice of lying on search warrants—a practice that amounted to a violation of the constitutional protection against unwarranted searches under the Fourth Amendment of the U.S. Constitution—had been exposed. Luna and his supervisor, Sergeant Hugo R. Amate, eventually admitted they had routinely made up informants as a way to cut corners. The two disgraced officers were convicted of perjury and lost their jobs.

Kenny Conley was a new cadet at a time when the Boston Police Department was under fire, when its long-calcified culture of lying and cover-up was spilling increasingly into public view. Commissioner Roache and his top brass found themselves on the defensive, insisting publicly the corruption was isolated. But the rank-and-file privately knew otherwise: Luna and Amate were scapegoats for a broader pattern of corruption. A diary entry by the chief of homicide, made public years later, reflected this. The chief noted the commissioner was angry about the fallout created by the drug unit's missteps. But, the cop noted, "no blame can be attached" to Luna and his supervisor because concocting fake informants and cutting corners was "the way the system operated.

"Because the acts of the drug officers imperil the police commissioner it appears that he is upset with us. If the case happened the same way tomorrow we would have to do the same thing. It looks like a case of wanting to shoot the messenger."

When Kenny and Bobby Dwan responded to their first call the early morning of January 25 and arrived at 36 East Newton Street, the street corner was barren. Nobody was around, never mind prostitutes. The durability of prostitutes always amazed Kenny—the idea they'd be out on a weeknight in 29-degree weather. "If they need the money they go out there in their little skirts, whatever," he said. On the other hand, the cold did put a chill into the level of illegal commercial activity. In that way, Mother Nature was an ally, an anti-crime initiative. Kenny and Bobby hung out for a bit, and then by one o'clock cleared the scene—calling in an "8-boy," police code for no persons found.

Kenny had become a full-fledged police officer after serving four years as a cadet. He had directed traffic during rush hour, he had worked in operations on a night shift answering 911 emergency calls, and, lastly, he had worked in the commissioner's office as a gofer. "Paperwork," Kenny said. "I was basically a secretary." On January 14, 1991, Kenny entered the police academy, and six months later, he was assigned to Area D–4.

Kenny was twenty-two years old. Graduation from the academy on June 19, 1991, was one of the most important days of his life. His family gathered for the ceremony, and Kenny proudly posed for photographs in his uniform with his mom and dad, and with Mike Doyle, who was also sworn in and was now on the force. Several nights later, Kenny, Mike Doyle, and other new officers hosted a celebration in Southie at the teachers' union hall. Father Toomey showed up to see his former crew and to congratulate Kenny and Mike.

Kenny was living at home on H Street. His father was gone, but his twin sister, Kris, was there, as she attended Emerson College in the Back Bay. His older sister, Cheryl, lived in the house with her two kids, too. Kenny no longer slept in his boyhood bedroom on the third floor. He made the basement into a makeshift bachelor's pad, laying down carpet and using a separate entrance in back.

Kenny worked hard—both his regular shift out of the Area D–4 station and details to earn extra money. In his free time he'd work out and play basketball in a couple of men's leagues, including Gatie's. For a few years he played football on Sundays. He hung out with the same friends from the corner of H and Fifth. They'd bring a beer cooler to the basketball games. Or, after games, they'd go home and shower and grab beers at The Cornerstone in Southie, which was owned by a family friend, or the Corner Tavern at K and Second Streets. Kenny's sandy-brown hair began thinning prematurely, and he became a "cap guy" with a growing collection of Red Sox, hockey, and other caps.

The mega-blow to the Conley family came in the fall of 1994. No one saw it coming. Maureen Conley had been working her shifts as a banquet waitress without incident. She had not had a drink in more than a year. "She was doing great," Kenny said. Kenny was out working a detail that October 19, 1994, Kris was leaving for work when her mother said she did not feel quite right. Their aunt came by and saw that Maureen was in trouble. Kenny was called and rushed home. Kris called an ambulance. Their mother was taken to the New England Medical Center, where doctors discovered she had acute kidney failure due to hepatorenal syndrome, along with liver disease.

"She went into a coma," said Kris. "She had surgery, but never recovered." Five and half weeks later, Maureen Conley was dead. "It was so sudden," said Kris. "Boom."

Two months later, Kenny was still shaken but kept his grief to himself.

Kenny and Bobby Dwan had barely cleared the East Newton Street area when they were called back, this time to talk to a man who lived in an apartment above a restaurant. The man complained nervously about the goings-on in apartment 3, saying drug dealers lived there. "He said there was supposed to be a drug shipment coming in," Kenny said. Intrigued, Kenny and Bobby stuck around to see if the man was right.

They began the stakeout at 1:09 A.M. By 2:07 A.M., they'd had enough. Bobby called the dispatcher and they pulled away. "Nothing happened," said Kenny.

Meanwhile, in another part of the city, Mike Cox and the gang unit had high hopes for the club Cortee's, where an assembly of hip-hoppers included Robert "Smut" Brown and his friends. For

Kenny Conley, though, the shift was shaping up like another ordinary night in what so far had been an ordinary career. In his four years on the force, Kenny had never been shot at. He'd never had to shoot at someone. He'd made plenty of arrests, but never a major one—such as a collar in a murder case. The absence of medals on his wall did not bother him. He was a young officer who did his job without fanfare.

But that was about to change, and despite what was in store for him, Kenny came up with a bit of gallows humor for that night. He would say years later that January 25, 1995, was one night "I wish I'd called in sick."

CHAPTER 4

The Troubled Boston PD

When Mike Cox and Kenny Conley were finding their footing during the early 1990s as neophytes on the force, the Boston Police Department itself was wracked by controversy. It seemed that with disturbing regularity a high-profile incident, involving tragic consequences, displayed the department's weaknesses, corruption, and an impenetrable us-versus-them mentality. The 1988 shooting death of Detective Sherman C. Griffiths during a drug raid exposed entrenched corruption in Griffiths's drug unit, where officers routinely and brazenly fabricated confidential sources to secure court-approved warrants. The department's reputation took a huge hit, as the fallout spilled over into the 1990s.

Then in late October 1989, a horrific crime in Boston captured the nation's attention—and, by the time it was over, showcased how the Boston police and the media had fallen prey to racial stereotypes in the worst way. In the early evening, a suburban couple leaving a birthing class at Brigham and Women's Hospital, one of Boston's premier hospitals, was shot in Roxbury while in their car. The husband, Charles Stuart, suffered gunshot wounds but survived. His wife, Carol, died within twenty-four hours, and their son, born by emergency Caesarean section, died seventeen days later. Charles Stuart claimed a black man had robbed and shot them. Within days,

Boston police stormed through a nearby housing project, turning it upside down and hunting for William Bennett, the black man homicide detectives insisted was the killer. The media coverage was unrelenting and swept the country. But the storyline turned out to be all wrong—in fact, it was a perverse and deadly hoax perpetrated by Charles Stuart. Nearly ten weeks after the sensational murders, it was revealed Stuart was the triggerman who shot and killed his wife. Charles Stuart committed suicide on January 3, 1990, by jumping off the Tobin Bridge into the Mystic River. Leaders of the minority community claimed Boston police were unable to see past racial blinders and had violated blacks' civil rights during its reckless manhunt in Roxbury. The finger-pointing, lawsuits, and repercussions lasted for months and months.

Later in 1990, a nineteen-year-old man was shot and killed by two Boston police officers after he'd shot four times at the officers. The teenager became the first of five people shot and killed by police during the next twelve months. Police practices soon came under in-depth press scrutiny, when in the spring of 1991 the *Boston Globe* published a four-part series about the Boston police titled "Bungling the Basics." Police officials were outraged and produced a point-by-point rebuttal. The city's mayor, Raymond L. Flynn, meanwhile announced in May the formation of a blue-ribbon committee to review the newspaper's findings. Flynn persuaded one of the country's best-known attorneys to chair the committee—James D. St. Clair, who in the 1970s had served as special counsel to President Richard M. Nixon during Watergate. In his thank-you letter to St. Clair for accepting the post, Flynn seemed to tip his hand—that he'd be happy to secure a clean bill of health for the police department. He noted in his letter that "Police Commissioner Roache has raised questions about the accuracy of the information contained in the article and the conclusions drawn by the reporter."

But if Mayor Flynn was looking for a whitewash, he didn't get one. Ten months later, on January 14, 1992, the "St. Clair Commission" submitted its findings to Flynn in a blistering 150-page report, concluding the police department's workings were deeply flawed and that Police Commissioner Mickey Roache was an utter failure who should step down.

"I've always believed if you have a talented team and it's not winning games, you fire the manager," St. Clair told the *Boston Globe* about the panel's recommendation that Flynn fire Roache. "This team has talent, but it's not winning any games."

The panel had interviewed hundreds of residents and more than eighty police officers. "We found poor morale among the police force and a growing impatience in the community," wrote the panel. "It is clear to us that most officers with whom we spoke and many segments of the community have lost confidence in Commissioner Roache and his command staff's ability to lead and manage the department."

The special commission pointed to the department's inability to police its own as a singular failure of wide-ranging impact that put residents in danger, fueled mistrust residents felt about police, and tarnished the department's reputation. "The failure to monitor and evaluate the performance of police officers—particularly those with established patterns of alleged misconduct—is a major deficiency," it said.

The panel conducted a painstaking audit of the department's Internal Affairs Division. The division's work, the panel found, featured, "shoddy, half-hearted investigations, lengthy delays and inadequate record-keeping and documentation." The panel discovered that less than 6 percent of all complaints of police misconduct filed by citizens were sustained as valid during the two-year period of 1989 and 1990. "This statistic strains the imagination," the panel said. "It assumes that more than 9 out of 10 citizens who complain of police misconduct are mistaken or are lying." The panel also found that a group of officers had gone largely unpunished even though they were repeat offenders and responsible for a "disturbing pattern of violence towards citizens." In short, the department was brushing off police wrongdoing, not rooting it out. For the rank-and-file officers, the reality was that misconduct was no big deal—

it rarely got them in trouble. Internal investigators either booted the investigation or did not look into the allegations at all.

In addition to calling for Commissioner Roache's removal, the St. Clair Commission recommended an overhaul of the Internal Affairs Division to include developing an "early warning system" to identify those officers with multiple misconduct complaints so that they could be targeted for retraining and even counseling.

The St. Clair Commission's findings were big news. It put Mayor Flynn back on his heels. In response, he said the department would implement many of the suggested reforms. For example, the department adopted a so-called Early Intervention System (EIS) to identify wayward officers. Initially, EIS required more than twenty complaints against an officer to trigger a review of the officer's conduct. The high threshold seemed a joke, more a throwback to the past than a reflection of forward-looking reform, and soon enough the threshold was lowered to three complaints within a two-year period.

But Flynn stood by Roache, the commissioner he'd appointed. Flynn and Roache were boyhood friends who'd grown up together in Southie. "Mickey Roache may not have done very well in the area of management," Flynn told reporters at a press conference. "But I give Mickey Roache an A-plus in the area of integrity and in the area of bringing people together racially." Roache hung on to his job for another year. When Flynn was appointed to be American ambassador to the Vatican in 1993, Roach resigned to run for mayor, and Bill Bratton stepped into the commissioner's seat.

Robert "Smut" Brown didn't have to read the St. Clair report to know the black community in Boston's inner-city neighborhoods deeply mistrusted the Boston police. Smut knew firsthand the kinds of abuses covered in the official report in clinical fashion. More than once, he said, cops had "done me wrong." Following one high-profile killing, he was leaving the Rose Club and about to get into his car when two officers snapped him up. "They cuffed me and put me in the car and rode me around, asking questions." He'd heard similar stories up and down the street. The alienation was embedded in the daily life of Roxbury, Dorchester, and Mattapan.

Black teens—some gangbangers, some not—had been complaining regularly to Boston and state officials about Boston police roughing them up. One girl, a seventeen-year-old tenth grader, said a police officer stopped her while she was walking down the street and asked her name. Why? she asked. The officer patted her down. When her two friends protested, the officer turned on them and patted them down too. In early 1990, a twenty-year-old black man said he was parked outside one of the city high schools waiting to pick up a friend when two officers walked up and began searching his car. He questioned the officers and said the officers told him to "Shut the fuck up." Young blacks said officers stopped them without reason—ordering them to drop their pants in public or to open their mouths for inspections or to place themselves spread-eagle against a wall.

For their part, Boston police were scrambling to combat the frightening rise in violent crime fueled by rampant drug use, especially crack cocaine. The 95 homicides in 1988 skyrocketed to 152 by 1990, with taped-off crime scenes becoming as frequent as sunup and sunset. Street gangs flourished, and some police commanders admitted openly that officers on the frontlines in Roxbury, Dorchester, and Mattapan were aggressively going after known gang members and anyone else associating with them. One captain in Roxbury was quoted in the *Boston Herald* in 1989 saying, "People are going to say we're violating their constitutional rights, but we're not too concerned about that . . . If we have to violate their rights—if that's what it takes—then that's what we're going to do."

The captain later called the remark a "psychological ploy" and insisted officers were not randomly harassing blacks or trampling on their rights. Police Commissioner Roache, soon after the captain's public comments, even issued a May 23, 1989, memorandum describing the department's "profound responsibility" to honor and protect citizens' rights under the U.S. Constitution and Massachusetts law.

Certain Superior Court judges were not convinced. One judge threw out two criminal cases after ruling Boston police had violated constitutional protections against unwarranted searches and pat downs. The famous U.S. Supreme Court case controlling these circumstances was a 1968 case known as *Terry v. Ohio*. In it, the nation's highest court ruled police cannot stop and frisk someone on a hunch, but must have a "reasonable suspicion" the person was engaged in or about to engage in criminal activity. In Boston, Superior Court Judge Cortland Mathers found that Boston police were violating so-called *Terry* principles by practicing a policy that "all known gang members and their associates (whether known to be gang members or not) would be searched on sight." The policy, the judge ruled, was tantamount to "a proclamation of martial law in Roxbury" against street gangs and other young blacks.

Judge Mathers was also unimpressed by Commissioner Roache's memo seeking to reassure the public that police were not running roughshod over people's rights. In a scathing rejection, Mathers said, "The Court finds a tacit understanding exists in the Boston Police Department that constitutionally impermissible searches will not only be countenanced but applauded in the Roxbury area." The police brass continued to disagree and defend the department in a war of words with their critics.

Newspaper headlines were one thing, the street was another. Smut Brown and others like him lived in a world where they believed Boston police would do anything to get them—lie, cheat, set you up. There was anecdotal evidence to draw on—instances where some officers didn't care about the truth and the law, only the conviction.

In one 1989 court case, an officer testified he watched a drug deal go down. But the officer would have needed X-ray vision for his testimony to have been true—he was standing on the other side of a two-story building when the alleged drug deal occurred. In a gun case, another officer testified he saw a gun stashed on a second-floor landing—and, again, this was superhuman. The officer would have had to be 30 feet tall for the eyewitness testimony to be true.

The police perjury had a name: "testilying." The fabrications might be rooted in good intentions—to convict the guilty. The officers wanted to arm prosecutors with the strongest, cleanest testimony possible, so they'd massage evidence against the accused to make the case seem better than, in fact, it was. They might honestly believe they had a criminal in their sights, and they wanted to make sure he didn't get away, even if it meant lying. They perverted the law to enforce the law. The ends justified the means.

It was nonetheless a message of injustice—and word about the bad cases got around. The worst examples were when the police department's freewheeling ways with the truth ended not with getting the right guy—even if by questionable means—but in a wrongful conviction. This happened in one of the biggest murder cases of the time. On an August night in 1988, a twelve-year-old girl named Tiffany Moore was perched on a blue mailbox on Humboldt Avenue, long the focal point in Roxbury's drug world and nicknamed "heroin alley." The broad street was scarred by burned-out homes, empty lots, and broken-down cars. Cash, clothes, and cocaine dominated its culture.

Tiffany was sitting on the mailbox swinging her legs, talking with friends. Then, from behind, two or three young men wearing Halloween masks ran across a small lot and began firing into the group. Minutes later, the 911 call to the Boston police captured the horror: "Oh, God! Oh, the little girl on the ground, shot." Blood poured from three bullet wounds. One—to Tiffany's head—was the wound a medical examiner termed "incompatible with life."

Tiffany Moore became an instant symbol of the drug-fueled lawlessness rocking the city. The girl was the youngest victim ever in the city's street gang wars, and her killing made the news around the country. She was collateral damage—the unintended victim of one street gang—Castlegate—seeking vengeance against the Humboldt Street gang, whose members were among the group of kids mingling on the street corner. City leaders sought to calm a public crying out for an arrest and panicked by the soaring murder rate. Some in the community even called for the deployment of the National Guard in Roxbury. Promising results, police launched a massive search.

Two tense weeks later, justice was apparently in hand. Shawn Drumgold, a twenty-two-year-old only a few months out of prison, and a second man were charged with killing Tiffany Moore. Police and prosecutors told reporters Drumgold was a "drug dealer and member of the Castlegate gang" and that "many, many witnesses" told them Drumgold was the shooter. The big problem with the statements was accuracy: They were false. Drumgold was no innocent—a street-corner drug dealer who had shot and been shot at, he surely fit the profile of a possible suspect. But Drumgold was a freelance drug dealer unaffiliated with any gang. Homicide detectives knew this; police kept books listing street gang members and anyone associated with a gang. It was all part of the beefed-up effort to combat the gang violence. Drumgold was not listed in the Castlegate book—or in any gang listing. Even some of the officers assigned to the streets of Roxbury were taken aback when homicide detectives picked up Drumgold as their man. "Shawn was dealing in peace, not bothering either gang," said one officer based in Roxbury.

But the homicide detectives apparently didn't want to hear any of it. From the start they focused on Drumgold and his pal, building a case on the backs of youngsters who were intimidated and pressured into providing incriminating testimony, all of which became pieces of the prosecution's case. Without key physical evidence—the guns and Halloween masks were never recovered—witness testimony was everything.

"I'm just a dumb puppet in there," one witness confessed later about how he folded under police pressure and agreed with his interrogators' suggestions that Drumgold was armed and looking for trouble that night. Fourteen months after Tiffany was killed, in October 1989, Drumgold was convicted of first-degree murder and sentenced to life without parole. His associate was acquitted. It didn't matter that many in the neighborhood knew Drumgold couldn't have done it, because he couldn't be two places at once. He was not on Humboldt Avenue when Tiffany was shot; he was blocks away with a group of friends snorting coke. But, given their illegal drug activity and fear of police, those alibi witnesses went underground rather than risk facing the police.

Even Tiffany Moore's mother was not sure. "It was very hard to tell if he was the right one," Alice Moore told a neighborhood newspaper, the *Bay State Banner*, following the murder trial in October 1989. "It's a big mess." Fourteen years later, Drumgold's conviction was overturned. Press accounts exposed the pattern of witness intimidation, possible prosecutorial misconduct, and the alibi evidence. In the "interests of justice," the district attorney asked a Superior Court judge to let Drumgold go home.

With the St. Clair Commission's explosive findings, 1992 was off to a terrible start for the Boston Police Department. But there was more trouble to come. That same January a trial began in Boston that served as a kind of appendix to the panel's findings—a case of police brutality dramatically illustrating in real life the failings the panel had outlined clinically and statistically.

The case was the May 1988 beating of the coke-snorting John L. Smith, who, after running a red light, had led police on a fifteen-minute chase through various city neighborhoods until his Cadillac broke down on Borland Street in Brookline. Smith's arrest woke up the residents, mostly professionals living on the usually quiet street not far from nearby Boston University. One woman told authorities she looked out her bedroom window and heard a "wailing sound" from Smith, who was lying facedown with his hands behind his back, apparently handcuffed. "It was a very eerie sound. It sounded like the sound coming from an injured animal. It was quite loud," she said. She watched officers casually walk over to stomp on Smith's back and then walk away. The woman's husband, a lawyer, was standing by her side at the window. He said he could hear Smith "whimpering and saying, 'No, no.' It was very loud, very clear and was obviously the voice of someone being hurt." To make matters worse for the police, another Borland Street resident happened to be a state prosecutor. Not just any state prosecutor either—Stephen L. Oleskey was supervisor of the Public Protection Bureau, which investigated complaints involving consumer protection and civil rights. When Oleskey looked out his window he first thought officers huddled around a man on the ground were trying to help someone injured in a car crash. When he stepped outside and police walked past him with Smith, he realized his initial assessment was likely wrong. "When I saw the civilian go by, with the battered face and handcuffed, I revised." Oleskey began thinking Smith might have been beaten. "But I wasn't sure."

Oleskey then talked to his neighbors. Before long the state attorney general's office opened an investigation. The next year it filed a civil rights lawsuit in state court against the police department, the city, and the "Brighton 13." Thirteen officers were accused of beating Smith or standing by and doing nothing to stop it. The January 1989 lawsuit was filed as a last resort—after state prosecutors had urged Boston police officials to discipline the officers but nothing was done. "When I became convinced that nothing was going to be done internally by the Boston police, I felt it was necessary to go to court," Massachusetts Attorney General James M. Shannon had told reporters.

It was a historic moment—state prosecutors asking the judge to issue a court order against the thirteen Boston officers barring them from using excessive force and demanding they report police brutality. Never before had such an injunction been sought against police—either in Massachusetts or, as far as anyone could tell, in the country.

The attorney general continued to try to negotiate with the officers, their union, and police officials, but they would have none of it. The thirteen officers denied wrongdoing, saying any force that was used was necessary to subdue a "coke-crazed madman." The police union attacked state prosecutors for meddling and second-guessing officers whose split-second decisions may mean the difference between life and death. Union lawyers called the court action outrageous. One argued, "If you start thinking twice about using force or your gun because of an injunction, your life is in danger." The police department's Internal Affairs probe, meanwhile, never got past "go." The case was simply on file, with no action.

"The office felt no choice but to go forward," one state prosecutor explained about taking the case to trial. "Our attempts to resolve the matter had been rejected over a long period of time and the City of Boston had never disciplined the officers." Testimony began the same week the St. Clair report came out—a nonjury, civil rights trial before Superior Court Judge Hiller B. Zobel.

Eight of the thirteen officers were from the police district known as Area D, with five officers working in the D–4 station in the South End. This was Kenny Conley's station, and January 1992 marked the seventh month Kenny had been on the street in uniform as a full-fledged officer.

"I didn't know most of these guys," Kenny said about the five officers going to court each day to attend the trial, as well as the other officers in his station. The politics and scrutiny of the department were swirling at levels far above his status as a twenty-three-year-old rookie. "I didn't talk to many people. I did my job, and I went home."

But the trial of the Brighton 13 was the talk of the police department—and especially in the D–4 station. Kenny overheard plenty. "The guys were just upset that there were injunctions getting put out against them." Newspaper coverage typically triggered the talk. "Lunch, dinner, or breakfast, or whatever, the guys would be sitting around writing reports, and someone would have a paper." That's when the complaints, commentary, and existential questions began: What the hell's going on?

It wasn't pretty. "Guys were saying it was fucking stupid." Most found the attorney general's case traitorous, going after Boston officers for doing their job: "Getting charged with arresting a guy." There was no debate about it—no opposing or cautionary point of view, where some other officer, sergeant, or supervisor was telling everyone to calm down. No one was making the point that brutality charges were serious matters and, if true, unacceptable. "Everybody was hostile towards it," Kenny said. That consensus extended "throughout the department."

The two-week trial featured some remarkable testimony. Ten officers took the witness stand and denied beating Smith or seeing him beaten. Two of them provided a rare and stunning peek inside their world. One officer testified that he would never report another officer's misconduct or wrongdoing. When asked why, the officer talked about the job's dangers and the need for brotherhood. The officer told the court he would not go against another officer because his own life depended on his colleagues.

The second officer, a veteran of more than three decades, unabashedly told the court that in all his years on the force, he had yet to see another officer commit so much as an infraction of the department's regulations. When asked how this could be, the officer did not answer he'd simply never seen it happen. He said: "I never see it."

The remark seemed a passage right out of the code of silence, or "blue wall," where cops stuck together at all costs—turning a blind eye to any misconduct or worse—where cops refused to "rat" on another officer. This was true even though it was a police officer's job to rat. Telling on people—ratting people out—went to the job's core, the sworn duty to uphold the law. To catch lawbreakers, an officer must testify—inform on criminals, either as an eyewitness or by relaying evidence others have provided. But when the lawbreaker was a fellow officer, this principle collided head-on into the "blue wall." It was all about bonds and ties—and ties that bind. With criminals, no such bond existed. With a fellow officer, it did—and so the culture's code was one of see no evil, hear no evil, and speak no evil.

The prosecutor cited the two officers' words as part of the "evidence of the code of silence; that is, not that they sit around a darkened candle in a darkened room and make a pledge, but what we have from that witness stand . . . a plethora of evidence that these officers lived by a code which they will not testify against another officer."

Judge Zobel wouldn't go there—he neither agreed with nor rejected the attorney general's claims of a police code of silence. Instead of drawing any broad conclusions about the police culture, he focused on the thirteen officers before him. In so doing, he hammered them. In a forty-page ruling issued on May 8, 1992, Zobel concluded that thirteen "frustrated, disgusted and angry Boston police officers allowed emotion to supplant their training." They violated Smith's civil rights, "from a shared belief that in staging his own version of *Smokey and the Bandit*, Smith had showered a platoon of Boston police officers with disobedience and disrespect, in the process endangering their safety and making fools of them all. Thus they set out to teach this scofflaw a lesson."

The judge did not rule specifically that any one officer had "kicked or kneed" Smith, but found that officers had "hurled themselves on top of Smith" after hauling him from his car and that "no law-enforcement reason justified the piling-on." Moreover, the judge said the officers' demeanor in court and their sworn testimony did not reveal "the slightest remorse or regret about *any* of that morning's events." "They don't get it," the judge wrote. "They do not understand how improperly they behaved on Borland Street.

"Being forceful does not mean using excessive force. The pressure that society puts on police to apprehend criminals and deter wrongdoing cannot justify a misuse of the physical power which society entrusts to every police officer." The judge then concluded that outside judicial intervention was necessary—and he granted the attorney general's bid to impose a permanent injunction against the Brighton 13.

"I have determined that because police officers are not likely to regulate police conduct, an outside sanction here is necessary for the public good."

The officers were angry. "For the first time, to my knowledge, a judge has issued an injunction against on-duty officers," said one of their lawyers. "It virtually renders it impossible for the officers to function on the streets, with this injunction hanging around their neck." They appealed, but Zobel's ruling was upheld by the state's highest court, the Supreme Judicial Court. The Brighton 13 were then ordered to undergo "extensive re-training" in civil rights, the use of force, and telling the truth.

For all the hullabaloo surrounding the trial, life inside the department for the thirteen officers barely skipped a beat. Throughout, they'd remained on the job. Some won promotions. From the inside, it was as if rather than losing the court case, they'd won. Kenny Conley, not one to pay close attention to the case's legal twists, reached that conclusion. Nothing seemed to have changed. While riding around on patrol during the remainder of 1992 and into the next year, he figured the police had prevailed. "I would assume they ruled in the police officers' favor because they're all walking the street."

CHAPTER 5

Mike's Early Police Career

The Boston Police Department is the oldest department in the United States. In the early 1800s, the city's growth on every front—land area, population, immigration, and commerce—gave rise to a more socially complex metropolis. The city was fast outgrowing its original town meeting form of government aimed at keeping power in the hands of many rather than concentrated in the hands of a few. The need for a professional, full-time force to replace constables reflected these changes, and in 1838 the state legislature voted that Boston should assemble a formal police department.

For more than a century and until the civil rights era, the force was largely a fraternity of white men. Then, in 1970, six black and two Hispanic men sued Boston and various state agencies in federal court after their applications at the Boston Police Department were rejected. The department's racial makeup illustrated what they were up against: Only about 3.6 percent of the force was black—in a city where blacks made up about 16.3 percent of the city's total population of 641,000 residents.

The men claimed that Boston's hiring and recruitment practices—and those at every police department in Massachusetts for that matter—were discriminatory. They argued that the civil service examination was biased and violated their constitutional rights under the Fourteenth Amendment. The federal judge agreed, ruling in *Castro v. Beecher* that the exams "were discriminatory against minorities which did not share the prevailing white culture." The bias may not have been intentional, the judge said, but it had nonetheless resulted in years of an "unconscious lopsidedness of the [police] recruitment." Determined to come up with a strong remedy that would have an immediate impact, the judge in 1973 approved a plan known as the Castro decree.

The plan required the creation of two pools of applicants: One group contained blacks and Hispanics, and the second group contained nonminority applicants. The applicants in each group were then ranked—with rankings affected by several factors. The most obvious was the exam score: the higher the score, the higher the ranking. Being a veteran of the armed services and being a relative of a public safety officer killed or injured in the line of duty were other factors that boosted rankings.

The two applicant pools were then merged into a single master list. The first name on the new list was the highest ranking minority applicant, and the second name was the highest ranking nonminority applicant. To fill openings, the police department went down the list—in effect, alternating between the two groups. The first opening went to a minority applicant, the second to a nonminority. The system was intended to "facilitate the appointment . . . of one minority policeman for each white policeman." No one disputed that the plan favored minority applicants, who, by virtue of merging the two pools, ended up ranked higher than a white applicant who had a higher score on the exam or was related to an officer killed in the line of duty. That was the whole point of the affirmative action plan—to use race to correct a gross imbalance in the racial makeup of the force. The goal was to achieve "rough parity," where the percentage of minorities on the Boston police force mirrored the percentage of minorities living in the city.

Mike Cox was a second grader at St. Mary's in Brookline when the Castro decree's quota system first went into effect in 1973. Fifteen years later he sat down to take the civil service exam under the revised system with the hope of becoming a police officer. As always, his older sisters had been looking out for him. Lillian not only showed Mike the recruitment ad in the newspaper, she got him the forms he needed to apply.

Lillian was responding to an interest Mike had had since he was a boy, even if he was uncertain whether he could ever measure up to the job. "It seemed like a very difficult job," Mike said. Self-doubts aside, the Castro decree had guaranteed a police job was no longer a long shot for a young black man. By 1988, the look of the police force had changed radically. Of the 1,908 officers, 365 were blacks and Hispanics—making up 20.7 percent of the department's total workforce.

Mike became excited about applying. "It was an opportunity to do something which I would probably learn to like, and I always had an admiration for law and the legal field." Mike was also impressed by the pay. "It wasn't bad either." Earning power was no small thing to him in his new role as provider. The year of 1988 was proving to be a big and hurried one—a personal trifecta covering marriage, family, and career.

Mike and Kimberly were married on June 25, 1988. Vince Johnson was his best man. Mike had just turned twenty-three and Kimberly was twenty-two. The wedding came just a few weeks after Kimberly's graduation from Spelman College with a bachelor of science degree in biology. The newlyweds immediately settled in Boston. Five months later, on November 14, Michael Cox Jr. was born. Even with all that, Kimberly was intent on juggling motherhood and her medical school ambitions.

In the fast makeover of his life, Mike's one piece of unfinished business was Providence College. Mike had not returned to Providence following his semester at Morehouse. He became preoccupied with Kimberly, their marriage plans, and starting a family, and he decided he could not start a police career and continue school. "I always knew I could go back to school," he said. "I didn't know if I necessarily would have another opportunity to go into the police academy." When Mike told Kimberly he was going to drop out and apply to the police department, she supported him. They agreed that Mike would complete college later and go on to pursue another goal: a law degree.

Mike passed the civil service exam, filled out the department's thick application, had a physical, and sailed through a series of interviews. The process included an assessment by a mental health professional who, studying the paperwork and noticing Kimberly's academic bona fides, joked, "What's it like being married to someone who's much, much smarter than you?" Mike laughed. "It's a matter of opinion," he retorted.

Soon Mike got the good word: He was in—an affirmative action hire in the new class of recruits that would begin six months of training at the Boston police academy on February 27, 1989. Vince Johnson was one old friend taken aback by Mike's career move. "I couldn't believe he was a cop," he said. "He never seemed the type." But one of Mike's former neighbors on Winthrop Street in Roxbury was not surprised. "The profession he chose was a good one for him," the elderly woman named Seleata said. "He cares about people and always has."

Following graduation from the police academy, Mike was ready for prime time—and he was assigned to a new station opened by the city to beef up police coverage in the high-crime neighborhoods of Roxbury, Dorchester, and Mattapan. The older station, known as Area B–2, was located in the heart of Roxbury in Dudley Square, just a few blocks from where Mike grew up. The new facility, known as Area B–3, was built a couple miles south on Blue Hill Avenue in the Mattapan neighborhood.

Mike began his probation working as a patrol officer during the day shift. A few months later, Mike jumped at the chance to work a more pro-active assignment. Cornell James asked the rookie to work with him. James, a black officer in his early thirties, was a veteran working nights in plainclothes. His family lived on Whiting Street, a cross street off the Coxes' Winthrop Street, and the two families knew each other well.

Mike was eager to join James. For a rookie still on probation, working in plainclothes was the fast track. But Mike quickly learned of the assignment's unique dangers. One night he and Cornell James went after a car thief. Listening to the radio, they heard the suspect had abandoned the car and was fleeing on foot. Mike ran from the cruiser and began heading down a street hoping to cut the suspect off. He was playing a hunch. "I was going to the point where the suspect was going to be at." He was dressed in street clothes. His badge was on a necklace around his neck, his police ID was in his wallet, and he wore his service belt. But the police identification was mostly concealed. To any bystander, Mike looked like a black man on the run.

Suddenly, a police cruiser raced by and cut directly in front of him. To avoid being hit, Mike leaped and landed on the hood of the car. He tumbled to the ground, and before he could stand up one officer had grabbed him by his shoulders while a second officer had him by the throat. "I couldn't talk. He was choking me." Mike was unable to explain who he was even if he'd wanted to. The first officer suddenly let go; he'd seen the police ID. The second officer, meanwhile, "stood there trying to choke me and threw me on top of the car." The first officer began yelling that Mike was a cop and then the second officer finally backed off. They left to chase the suspect.

Mike was left dazed by the hit-and-run. "I wasn't scared," he said. "I was baffled." It was the choking by the second officer—a textbook example of the use of unnecessary force. "I didn't understand why he was choking me because I was offering no resistance, so I was more or less angry. I didn't know what was going on."

But Mike quickly let go of his concerns. After the suspect was apprehended, the two officers swung by the Area B–3 station to look him up. They were not from Mike's station and didn't know the rookie. "They all came by to apologize," Mike said. The first came up and said, "I didn't recognize you." The second also said he regretted the mistake. It ended there. But not before word got around and a captain in Mike's station called him into his office. "I gave an oral report," Mike said. The captain explained his concern: He'd heard talk the officers involved had a "reputation for doing things like that to black people who live in that area." He wondered if Mike wanted to file a formal complaint. No way, said Mike. He was a twenty-four-year-old rookie cop. "I said, 'Captain, I'm on probation. I just started this job. They all apologized. I'm satisfied with that.'"

When his probation ended early in 1990, Mike was assigned to stay in B–3 in Mattapan, although his shift changed to the "last half" from 11:45 P.M. to 7:30 in the morning. He started out patrolling alone in a service car, responding to routine calls from the dispatcher. Then one night, Craig Jones, another black officer at the station, asked Mike to team up with him. Craig was working in plainclothes in an anti-crime car—in fact, by this time, Craig and his partner were the only ones in that capacity in B–3. But Craig's partner was getting transferred downtown, and Craig needed a new partner.

The two didn't really know each other, although they'd seen each other at the station house. Craig had been on the force only a little more than two years. He grew up in Boston and, as part of busing, attended South Boston High School. Then he enlisted in the U.S. Army and served three years—at bases in Fort Lee, Virginia; Fort Knox, Kentucky; and Karlsruhe, Germany. When he was discharged, he worked as a security guard in a mall. He entered the police academy in late 1986 and hit the streets in 1987.

Mike took to Craig's offer instantly. He had liked working with Cornell James. He saw working in street clothes as carrying a bit of prestige along with the freedom to be really active on the street, the kind of police work he wanted: "Drug arrests, gangs, things like that." The two teamed up and, in short order, established themselves in Mattapan as a pair of enterprising crime-fighting cops drawn to the action. "We would go to places where they were known to sell drugs," Mike said. "Or there were a lot of shootings that we would respond to. Priority 1 calls only—the highest priority—it would be shootings or stabbings and gang calls."

One early morning in June, the new team of Cox and Jones played a key role in apprehending a man who'd shot another officer in the thigh. It was 2:30 A.M. on June 28—three days after Mike's second wedding anniversary—and the two raced to a housing project called Bataan Court after hearing about a shooting on their radio. They arrived in time to see a black man, brandishing a rifle, jump into a Pontiac. The ambulance was arriving to attend to the fallen officer, and Mike and Craig took off after the fleeing Pontiac. They were able to radio in the car's location to the dispatcher, and, several blocks away, another unit cut it off. Two men with rifles were arrested.

Six months later, at 4:15 A.M. in the morning of December 6, a convict who'd just gotten out of prison went on a shooting rampage in Dorchester—firing an AK–47 semiautomatic machine gun. Two residents were wounded, and a man was killed. Mike and Craig were one block away. They arrived to find two men standing in the middle of the street. The men began to run. Despite a mismatch in firepower, Mike and Craig jumped out and gave chase. One suspect stumbled, and the AK–47 hit the ground with a thud. The suspect kept going. Craig grabbed the weapon while Mike followed the man into a backyard farther down the street. Another officer, coming from the opposite direction, cut off the suspect and captured him. Mike's eyewitness account later convicted the man.

The next month, Mike and Craig were in their cruiser at 3:40 A.M., staking out a party when they saw two men leave the apartment building in a hurry. The men climbed into a car. Mike and Craig then heard on the radio that a shooting had occurred in the apartment. The dispatcher was calling for police units as well as an ambulance to respond. Mike and Craig were all over it. They turned on their lights, raced after the car, and cut it off. They arrested the two men inside and recovered a .38-caliber handgun from the car. The weapon was loaded with four live rounds and one spent shell.

Time and again, Mike and Craig thrust themselves into the thick of it. In the process they'd learned about each other. Both were tall and strong and took pride in their physical fitness. They discovered they complemented each other personality-wise—a bit of yin and yang. Craig ran hot—he was typically a step ahead at an incident, jumping in to size up the crime. "He liked doing that, you know," said Mike. "Going inside, see what's going on." Mike ran cooler and quieter. He tended to work the perimeter.

It worked. They were a good fit. They made arrests—or assisted in arrests—that were clean, intense, and exciting. High-five police moments. Barely two years on the force, and Mike and Craig shared awards for exemplary police work. Their role in arresting the man who shot another officer, for example, resulted in a "medal of honor" in 1991. Then, in early 1993, they won a promotion to the elite gang unit. By January 25, 1995, Mike, having done well on the exam, was awaiting a promotion to sergeant.

Along the way, however, were nights when the crime fighting was not so clean, when in the heat of the moment the lines between the good guys and the bad guys became confused and complicated, when Mike Cox experienced déjà vu to the first time he was mistaken as a suspect. One episode was later during Mike's rookie year, after he had completed his probation. He and several other officers were chasing a suspect on foot down a street in Mattapan. The suspect jumped over a fence. Mike ran along the fence to keep up with the suspect, and then he began climbing over it too. But as he climbed he felt someone grabbing at his legs, trying to pull him down. Mike turned and saw two officers. "There was apparently some mistaken identity," Mike said, but the confusion was short-lived. "I was able to verbally say who I was, and that more or less ended the physical grabbing of me." Mike wasn't troubled by it. "I wasn't punched or anything."

Nor was he troubled by several other incidents. Hearing a report one night that an armed man was walking in back of a building, Mike and Craig headed over to investigate. They carefully made their way down an alley when they noticed an officer they knew standing in the dark. They assumed the officer recognized them. But he had not. Mike and Craig, dressed in hooded sweatshirts like a pair of gangbangers, continued walking toward the officer. The officer shouted, "Show me your hands, show me your hands!" He drew his gun. Finally Craig said something. "Mark, Mark—it's me, Craig."

The officer relaxed, and the suspicious persons incident was recorded officially as an "8-boy," police code for no person to be found. Unreported was a mix-up that, in a blink, could have gone bad but luckily had not. For their part, Mike and Craig had not helped matters. They had not radioed ahead to say they were responding so that the officer in the alley would be on notice that two officers in plainclothes were on their way. But despite its dark potential, the moment passed—no harm, no foul.

Another night, Mike and Craig were walking through a housing project when they saw some officers searching a couple of suspects. One officer looked over and shouted, "Who the fuck are you?" He had not seen that Mike and Craig were police officers. But the confusion was cleared up before anything bad happened. Again, no harm, no foul.

Even so, the mistakes left their mark, a challenge to the idealism—or naivete—Mike brought to being a police officer. "When I first came on the job, I never really considered myself just a black police officer. I just considered myself a police officer." The words were classic Mike Cox: the Roxbury boy who'd gone to schools that were mostly white, from St. Mary's in Brookline to the private preparatory school Milton Academy, and then to boarding at the Wooster School in Connecticut. He was the young man from a middle-class black family who believed character and hard work meant more than race. In many ways, he was color-blind. "It was the way I was brought up," he said.

But race did matter. The instances of mistaken identity revealed the complexity of a more racially diverse police department. Mistakes like this didn't happen when the force was nearly all-white. The scare of mistaken identity was turning out to be a dangerous side effect of the department's successful affirmative action program as well as its initiative to fight the escalating street gang violence by sending black officers into the fray dressed in street clothes.

In time, Mike began to catch on to the risks that were race-based. "I realized that the job itself, it's a lot more dangerous, just because of the fact I'm black." But while more aware, Mike still wasn't too worried—a go-along attitude that wasn't shaken even after the worst mix-up right before he joined the gang unit.

Mike was running after a suspect down a residential street in Mattapan one rainy night. He was way ahead of a number of other officers, including Craig. The suspect was believed to be carrying a weapon. Mike held his handgun in his right hand as he pumped his arms trying to catch up. Suddenly an unmarked police cruiser pulled up alongside him. "I was running straight," Mike said, "and it drove alongside of me, and then it turned into me." Hit hard from the side, Mike was airborne. He slammed into a fence along the side of the road. The cruiser jumped the curb and kept after him. "It pinned me actually against the fence, so my feet were not on the ground." He didn't know what was going to happen next. "I was scared because I didn't know if the car was going to continue to run over me." He didn't feel any pain—yet—only fear. "I was very worried about being killed." Then the cruiser backed off, and Mike heard Craig's voice in the distance. "What the fuck are you guys doing?" he screamed. Craig ran up shouting at the officers, "Are you guys stupid! He's a cop! What the fuck!"

Mike was slumped on the ground, his left knee throbbing. More officers arrived and attended to him. Within minutes, he was taken by an ambulance to Brigham and Women's Hospital, where his badly swollen and cut knees were treated.

The injuries did not prove to be serious. Soon Mike was back at work. But nothing came of it officially. The police report required whenever an officer is hurt in the line of duty referred to Mike's injuries as unintentional: "Officer Michael Cox, along with other officers, was chasing a suspect armed with a handgun. During the foot pursuit Officer Cox was accidentally struck by an unmarked police cruiser." The report flatly contradicted Mike's view that the cruiser struck him intentionally.

"I voiced my displeasure after the fact," he said. But he did not file a formal complaint. Even though it was no longer no harm, no foul, Mike let it go. It was not his personality to be outspoken—and it never had been, going back to grade school when teachers worried about his reticence. Then at Wooster, a student body that featured the likes of Tracy Chapman was fairly active in social and political causes, but Mike was not. He just wanted the ball to drive to the hoop.

Mike preferred to brush aside the close call. He'd taken to police work and liked the feeling that he and Craig were having an impact on the streets. "I loved the job, and it was more than what I expected," he said. Mike could almost sound golly-gee about it. The job, he said, "meant integrity. It meant—gosh, respect. Loyalty to your community, and just to the people around you. It meant a lot of things to me."

Sure, the work could be a grind. He wasn't wild about the paperwork. And while others enjoyed the pageantry that might accompany working a parade or major city function, Mike did not. "You are just standing there in uniform, either in the heat or the cold, not really doing anything for long periods of time. It was just tedious." And he did not enjoy appearing in court to testify in a case. "Just answering ridiculous questions," he said. Mike viewed those as chores getting in the way of what mattered most. "The actual police work. Going out and thinking that I was making a difference somehow, you know, by arresting people who were truly bad and helping people who really, really needed help. Those two aspects were the two things I liked the most."

His was a truly satisfying start to a career—especially for someone who'd dreamed of becoming a police officer but was insecure whether he was good enough to serve. "I really had these high standards for what police officers were," he said. "I didn't know if I really necessarily measured up to those standards." Very quickly Mike had learned he did measure up. "I took a lot of pride in it. I don't want to say I got self-esteem from it, but what I did was certainly something I was proud of."

His wife saw it too. "I was a little nervous in the beginning," Kimberly said. But she realized her husband had indeed made the right choice. "When he was going through basic training, you know, he loved the friendships that he made, the people he met."

Kimberly had hit on something. The brotherhood aspect of policing was a big part of Mike's satisfaction with the job. Mike and Craig Jones were more than friends, for example. They experienced a deep trust from working side-by-side. "When you're in dangerous situations and you work in dangerous places, you have to have a certain bond," Mike said. "Just to go in and out of those places and feel comfortable, knowing you're going to come out okay, in the sense of watching your back." Their closeness was rooted in the very ethos of big-city policing—the us-versus-them mentality, where it was the law against the lawless. Mike embraced the loyalty flowing both ways.

And he enjoyed the ebb and flow of their nights, from the intensity of a crime scene to the relaxed banter back in the office. During a shift, Mike and Craig often grabbed a bite in the cafeteria of Carney Hospital in Dorchester just outside their Mattapan district. They'd meet up with other officers working through the night. One was David C. Williams, a black officer working in Dorchester in Area C–11. Williams, born in Trinidad, had moved to Boston when he was nine and grew up in Dorchester in Uphams Corner. He'd been a police officer for almost four years, joining in 1991, or two years after Mike. There was Richie Walker, a black officer who wore his hair in braids and worked in Mattapan in the area known as B–3. Of the group, Walker had been on the force the longest—since 1985—although his ten-year run was interrupted when, while off duty, he'd pulled a gun on a civilian after a traffic accident. Walker was actually fired, but he appealed the dismissal through labor arbitration and won his job back. All in all, the eating club was a chance to connect in a setting more relaxed than a crime scene.

For Mike and Kimberly, home life was nothing if not hectic. Mike was still a rookie cop when Kimberly began commuting to Philadelphia in the fall of 1989 to attend medical school. Mike Jr. was ten months old and she was expecting their second, but she was determined to get her medical studies under way. Nicholas Cox was born on January 14, 1990, in Boston, and Kimberly went on a leave for the remainder of the academic year. She resumed her studies the following September 1990, spending weekdays in Philadelphia and commuting home to Boston on the weekend. Mike cared for the boys during the day, and when he left for work at night, his mother, Bertha, usually stayed with the two babies, who were fourteen months apart.

The couple made it work, no matter how stressful their lifestyle. "Being away," said Kimberly, "commuting back and forth, having two small kids, trying to get through medical school, and Mike trying to support all of us." The challenge brought them closer. "We didn't have really any major disagreements," Kimberly said. "Basically, I was in school and he was taking care of the kids, and I trusted him and he did an excellent job and he managed and handled everything." She valued Mike's soft-spoken way, his steadiness and levelheadedness. She loved that Mike was a "nice, easygoing person who enjoyed doing things with his family," and she loved how Mike made her laugh. "He would joke a lot. He had fun. We did things together." They imagined a future when she was a doctor and he was a police officer armed with a law degree. "We were looking forward to this new and wonderful life together," Kimberly said.

Mike was especially proud of providing for his family. In five years he'd doubled his earnings. He started out making $30,115.53, including overtime. His earnings jumped to $61,394.82 in 1994. The couple watched their spending and saved money by moving into the two-family house on Supple Road owned by Mike's oldest sister. Cora L. Davis, eighteen years older than Mike, lived upstairs with her husband and kids. The couple's future was bright. Mike even talked about taking a class here and there to finish college.

Simmering beneath the surface of Mike's police work, however, was the problem of black police officers being mistaken as suspects. The department had not caught up to the vexing aspects of a police force with increasing numbers of black officers. There was no proven method for an officer in street clothes to signal other officers that he was one of them. Most of the time Mike and other black officers simply relied on being recognized.

Mike did once speak up about the problem. He and Craig, brand-new to the anti-gang unit, happened to cross paths with the newly installed police commissioner one night in July 1993. William Bratton, flamboyant and ambitious, had joined the Boston force in 1970 and then left to hold leadership positions with several different police agencies. Most recently he'd been chief of the New York City Transit Police. The new chief, two weeks on the job as Boston's top cop and saying he wanted to check out the front lines himself, went out riding with a patrol officer in Mattapan. At 10:30 P.M., a dispatcher put out a call about shots fired nearby. Mike and Craig were the first to get there. They pulled up to a group of young black men. One kid turned and bolted. Mike called for backup while Craig ran and captured the fourteen-year-old suspect.

Bratton then arrived. He got out, looking around, and under a fence found something the fleeing teen had dropped—a silver .32-caliber revolver. The boy was arrested on gun charges. Despite the late hour, a *Boston Globe* reporter learned about Bratton's hands-on police work and wrote the kind of flattering item the press-savvy Bratton cherished. "Police Commissioner Spends Night on Duty" was the story's headline, and it quoted Craig. "He knows what he's doing," he said about Bratton.

Bratton switched cars to ride with Mike and Craig. Neither knew Bratton. The commissioner, who had a reputation as a progressive, hard-driving administrator, was the one who raised the subject of race and working in street clothes.

"He mentioned it," Mike said. "He asked, had we ever been mistaken for suspects before by other officers and felt as though our life was in danger by the other officers."

Mike and Craig both answered they had.

"We started to tell him some of the examples," Mike said.

Bratton said he'd seen the same problem in New York City. He then asked Mike and Craig whether they thought the Boston Police Department should develop "some type of system" to identify officers working in street clothes, particularly black officers.

Mike and Craig both answered yes.

The ride-along ended. Bratton thanked them. The two gang unit officers felt the unexpected exchange with the new commissioner had gone well. It was good talk.

But no action followed. Within a year, Bratton was on the road again, this time to become chief of the New York City Police Department. "He didn't implement anything," Mike said.

CHAPTER 6

Closing Time at the Cortee's

At the club Cortee's, the fact that Smut and Mike and Craig did not bump into one another was simply one of life's happenstances. By the time Smut had arrived around midnight, Mike and Craig had completed their quick turn inside to gauge Hip-Hop Night and were already riding off in their Tango K–8 car.

It wasn't as if they didn't know one another; Mike and Craig had played cat and mouse with Smut ever since Smut got home from jail in 1992. By then, the two cops in plainclothes were known on the street for pulling up fast in their unmarked cruiser and jumping out to confront gatherings of "hoodies." The in-your-face arrival was not solely cop macho; it had a purpose. "On the street the way these guys work is by intimidation," Mike said, "so jumping out showed them we're not intimidated."

The up-tempo entry was also a barometer, a way to gauge a street gang's level of current criminal activity. If the kids reacted with swagger and trash talk, then they were likely just hanging out, not up to any trouble. But if the kids went silent, looked away, or tried to melt into the night, "then we'd sense something was up. Something was just finished or something was in the works."

Mattie Brown, however, was not impressed with their headstrong style. She nicknamed them the "Jump-out Boys." Her son and his friends often hung out in front of her house, and then along came Cox and Jones. "I'd yell at them from my porch," she said. "Cuss 'em out to get off my property. Sometimes they'd yell back, 'The street isn't private property! We can do what we want.'"

For his part, Smut regarded the two cops as no-nonsense, but straight-up and honest. The two cops, meanwhile, saw Smut as principally a dealer who seemed levelheaded, "one of the more reasonable ones in the group," Mike said. Mike and Craig were mainly after gangbangers who specialized in the lethal combination of drugs and guns. Smut Brown, said Mike, "was not a shooter, not a gun guy."

Of course, that didn't mean Mike and Craig were going to look the other way. The "Jump-out Boys" and Smut did eventually have a memorable clash. Mike and Craig were heading down West Selden Street one night. It was about eleven o'clock on March 23, 1993. Smut was driving up West Selden—fast—and he roared past the two cops. Mike guessed Smut was hitting about 50 mph, well above the speed limit for the residential neighborhood. He and Craig turned. Smut cut sharply down a side street and then turned onto another street running parallel to West Selden. Mike and Craig caught up, the lights on their cruiser flashing. Smut pulled over to the curb and jumped out. He ran across the street and was heading between two parked cars. Mike followed and caught up to Smut on the sidewalk near the parked cars. Craig joined them. Something on the ground between the parked cars caught Mike's eye. He picked up a plastic bag. Inside was Smut's stash—sixteen pieces of crack cocaine individually wrapped and ready for sale.

Four weeks later, Smut was found guilty of possession of cocaine with the intent to distribute by a judge in Dorchester District Court. The judge sentenced him to serve a year in the House of Corrections. Smut immediately appealed. His lawyer and the prosecutor then worked out a deal. Smut would plead guilty to the lesser charge of coke possession, and, in exchange for the guilty plea and dropping the appeal, the jail term would be suspended. Smut was placed on probation for two years—or until July 1995.

Once again, he'd managed to stay on the street. The stay-out-of-jail card meant when Smut walked into the club Cortee's on the night of January 24, 1995, he was on probation from the Mike Cox bust.

Once he was inside, Smut never noticed that in back, Lyle Jackson was seated at a table playing cards, dressed in Boss blue jeans and a purple-colored Champion sweatshirt. Years had passed since Smut and Mama Janet's son played together, either at Lyle's house off Humboldt Avenue or at Smut's Franklin Hill housing project.

Lyle quit the card game just after midnight and hooked up with a couple of his friends, one named Marcello and the other named Stanley. They drank and mingled around for a while, and by 1:30 or so, Lyle and Marcello decided they were hungry. One of them mentioned Walaikum's, a tiny hamburger joint about a half mile away on Blue Hill Avenue in the Grove Hall section of Roxbury. Of the three, Stanley was the one with a car, a red Hyundai, so they asked him for a ride. Stanley said fine; he was hungry too.

Like Lyle and his friends, a lot of people at the club were starting to pull out, including Smut and his crew of Tiny, Tiny's brother Marquis, and Boogie-Down. Closing time was approaching, and the club had stopped serving drinks. The music was winding down. Tiny was still on edge after seeing Little Greg. He might have recovered from the leg wound from the summer before, but there was no recovering from their beef. Smut tried his best to keep Tiny distracted. "I told him to chill." The four discussed a nightcap. Smut suggested they head down to Mattapan where he knew an after-hours place. Tiny liked the idea because his mother lived in Mattapan and he wanted to swing by her house.

Meanwhile, Mike and Craig were outside turning their car around to resume their surveillance position on the hill overlooking the club. During the dustup with the girls, other members of the gang unit had not blown their cover. Donald Caisey, for one, was still in the decoy cab right on the street across from the club, while Joe Teahan and Gary Ryan were not far away in their unmarked cruiser.

But they were not the only cops waiting for something to happen. Unknown to the gang unit, another cop was nearby. Dave Williams also knew Hip-Hop Night at the Cortee's might get interesting. On a typical night, his Dorchester station's "batting order" for patrolling the district included four or five "service units" manned by a single officer, another two or three "rapid-response" units manned by two officers, and usually one "anti-crime" unit manned by two officers dressed in street clothes. Williams had begun the night working alone in a service unit known as the Harry 411.

Williams was assigned to patrol the Savin Hill neighborhood, a sector located on the east side of Dorchester, bordering Boston Harbor. It was several miles from the Cortee's on Washington Street—basically from one side of the district to the other. But Williams was nonchalant about straying so far from the patrol sector his shift sergeant had assigned. "They pretty much tell you, you have the Savin Hill area, but you can go anywhere you want," he said later. No one was really looking over his shoulder.

On the force, Williams was known as a "working cop." It was a label to distinguish cops like him from those seeking uneventful shifts that might even include a nap. Williams was action-oriented—so much so that he'd actually drawn some supervisory concern about a tendency toward "physical abuse during arrests." In fact, along these lines, Williams had had a rough few months. In September, a Dorchester woman complained to Internal Affairs Williams punched her out. She'd saved a clump of hair she claimed Williams had pulled from her scalp. The charge stemmed from a confrontation involving police and partygoers one Sunday morning over an illegally parked car. Williams admitted he hit the woman, but said he did so only after she and two of her friends attacked him. He'd met force with the minimum force necessary to subdue them, making the punch justifiable.

The next month his response to an "excessive noise" call led to another brutality complaint. Williams was one of three Boston police officers arriving at about 1:30 A.M. to a party in a third-floor apartment. After talking to the party's host, Williams walked past four young men hanging out on the front porch. One was a teenager named Valdir Fernandes, a seventeen-year-old high school student. As Williams walked down the steps, Valdir spat. Williams turned and demanded to know if the spit was aimed at him. Valdir denied any such thing. Valdir said later that Williams then bounded up the steps, pushed him against a wall, and "grabbed me by my throat, smacked me on the right side of the face." Valdir's mother rushed outside and confronted Williams. Williams insisted the boy had been disrespectful. He denied striking him at all. Valdir was taken by ambulance to Boston City Hospital, where he was treated for head trauma and for cuts and marks on the tracheal area of his neck. The family photographed the teenager's battered face and soon after filed a complaint against Williams with Internal Affairs.

That made two abuse complaints against Williams in two months—or a total of four in just over two years. In theory, the two recent complaints should have triggered the department's Early Intervention System, created in the early 1990s as part of a major reform effort. The intervention system was supposed to kick into gear when an officer received three complaints within a twenty-four-month period. But theory was one thing, the practices of the police department were another. By the night of January 24, four months had passed since the Valdir Fernandes incident and Williams hadn't heard a word about any intervention or retraining requirements. For the officer the brutality complaints were more a nuisance than any real threat to his standing in the department.

Having abandoned his assigned patrol, Williams was parked alone in his cruiser down the street from the Cortee's. He knew the club's reputation for trouble; the club, he said, was a place "where gang members were going and they were having fights and shots were fired." Initially he'd driven up a hill with the idea of watching the club from there, but then he spotted a gang unit vehicle. The unmarked car—which was Cox and Jones's—was sitting in a driveway. He didn't want to get in the way of an ongoing operation so he'd decided to keep moving. He'd driven down the street a few blocks past the club and pulled into an empty lot. He was out of sight and had his radio going.

"I just sat back," he said. "I think I was reading the paper."

Mike and Craig were back to their lookout on the hill when Lyle Jackson and Marcello climbed into Stanley's car to head over to Walaikum's. It was just before 2 A.M. Moments later, Smut Brown, Boogie-Down, Tiny, and Marquis left the club. Smut and Boogie-Down walked toward Smut's Volkswagen Fox around the corner. Smut had told the others the bar to go to was Conway's, where he knew the manager. He told Tiny to follow him. Tiny and Marquis walked toward the parking lot across from the club where Tiny had parked a 1994 gold Lexus, a "loaner" from the dealer. Tiny had bought a brand new GMC Jimmy SUV—a birthday present to himself—but the truck wasn't ready and the dealer had given him the Lexus to drive for a few days.

Smut pulled his car around the corner and was facing the club so Tiny and Marquis could see him. He and Boogie-Down sat there with the engine running and the heat cranked up against the sub-freezing cold. Smut noticed Tiny walking quickly toward them. He could see Tiny was agitated, moving in a jerking motion. Smut opened the window, and Tiny was stuttering about Little Greg, saying Little Greg was up to no good. Tiny pointed to the far side of the Cortee's entrance, where Smut saw a group of Castlegate gang members. Tiny talked nonstop, explaining he and Marquis had noticed the group before they got to the Lexus. But when he couldn't pick out Little Greg, he got suspicious. Where was Little Greg? Where was he? Tiny believed Little Greg was out to get him and planning to do something.

Tiny had given Marquis the car keys and gone to find Smut and Boogie-Down, all the while looking over his shoulder. He wanted protection. He ordered Boogie-Down to give him his gun. Boogie-Down resisted, but Tiny was adamant. He needed the gun. Boogie-Down handed it over, and Tiny turned to head back toward the club.

He was maybe two car lengths away when Smut detected a car

moving slowly past his from behind. Smut saw the flame of gunfire first. Then he heard the pop-pop of gunshots. Tiny jumped sideways behind a parked car. Tiny was not hit. One bullet struck the black Isuzu Rodeo in front of Smut's car, leaving a hole in the front door. The others went off into the night. Tiny stood up, gun raised high over his head, and he got off a few shots. The car sped up, driving past the club and up the street. Smut threw his Volkswagen into reverse. He backed into the intersection so he could turn the car around and drive away in the opposite direction. Tiny was running to the gold Lexus and jumped in so he and Marquis could follow.

Within seconds, the gang unit police radios exploded in noise. Voices collided.

"Shots fired! Shots fired!"

It sounded like the percussion section of an orchestra gone haywire.

"Shots fired! Right out front!"

Up on the hill, Mike and Craig exchanged looks of extreme frustration. "I'm like, Awww!" Mike said later. They'd just left the club and now the main event had started. This was the whole point of the night—guns and street gangs—and they weren't there.

They raced back and found the street crowded with people running in different directions. Some were jumping into cars while other cars were already pulling away, including the Volkswagen Fox and the gold Lexus. Mike saw the other guys from the unit looking around and trying to talk to people. Donald Caisey was there. The unit's supervisor, Sergeant Ike Thomas, was there. Teahan and his partner, Ryan.

"Everyone who was working in the gang unit that night," said Mike.

The gunshots also drew other officers working in the area. One was Richie Walker, the officer known for wearing his hair in braids, who worked out of the Mattapan station. Walker activated his siren and lights and immediately began heading toward the club. But on his way he saw a Peugeot speeding toward him and then turn abruptly down a side street. His participation in the main event would have to wait. Walker chased the car and arrested the twenty-one-year-old driver. He found plastic bags in the front seat containing "vegetable matter," or marijuana. The suspect was taken in by another officer, and Walker stayed with the Peugeot to await a tow truck. While he waited, he monitored the radio for updates on the shootings at the Cortee's.

One officer who did make it to the club was Jimmy Burgio, who, like Dave Williams, worked in Dorchester out of the C–11 station. Burgio was from Southie, a sports jock. He was crazy about ice hockey and made up for a lack of skating finesse with a bullet-hard slapshot launched from his muscular, husky frame. Burgio was starting his fifth year on the force and brought his competitiveness to the job. For example, he was extremely proud of the seven binders he'd filled with photos and intelligence about street gangs and criminals. He was wary, however, about sharing the intelligence—especially with the gang unit—for fear his enterprising work would get ripped off and he wouldn't get any credit.

He also viewed each night's shift as a full contact sport between the good guys and the bad guys. He talked about police work in these terms—a high-stakes, hard-checking rivalry played out nightly. Some nights, he won. Some nights, the bad guys won. He got charged up looking ahead to work—each shift being a competition with the suspense of not knowing who was going to come out on top. He loved the cop life.

No surprise that, like Dave Williams, Jimmy Burgio was known as a "working cop." No surprise either that, like Williams, Burgio strayed from his assigned patrol district when he heard the screaming voices on his radio about shots fired outside the Cortee's. Burgio raced to the scene. But once he saw plenty of police were already there he headed up the nearby hill to watch, taking up the overlook Mike and Craig had abandoned.

Burgio could see a combination of unmarked units used by the gang unit as well as marked cruisers. Blue lights flashing. Cops in uniform and in street clothes milling about. Young people were scattering quickly into the night. Burgio recognized some of the cops. He knew the gang unit team of Joe Teahan and Gary Ryan. He was recently engaged but had once dated the woman who ended up marrying Ryan. He knew Donald Caisey from work. Burgio did not know Mike Cox, but he did know Mike's partner, Craig. He had little use for Craig, considering him "an arrogant bastard," a glory hound.

When they got back to the Cortee's, Mike and Craig jumped from their cruiser to help the others canvass the scene. Mike was wearing the three-quarter-length black parka he'd borrowed from his nephew over a black hooded sweatshirt. He'd left the wool Oakland Raiders skullcap in the cruiser. It turned out the gunfire happened in a dead zone—a spot along the street where no officer saw what exactly happened. Mike and Craig tried talking to the patrons leaving the club, asking what they saw, who fired the shots. They studied faces, looking, said Craig, for "someone who looks suspicious." At one point, Mike spotted Dave Williams, and Dave spotted him. They knew each other pretty well and ate together sometimes at Carney Hospital. They exchanged a quick greeting.

No one was having any luck getting a lead on the shooters. The club's patrons were looking to avoid the police, not talk to them. "No victims came forward. No witnesses came forward," noted the spare police incident filed later about the shooting.

"Everybody just kind of got in their cars and left," Craig said later. Within minutes, the cops found themselves standing there looking at each other. It was quiet in an eerie sort of way.

The street was empty and the gang unit was empty-handed.

"We didn't get anybody," Mike said. "People were just pretty mad."

CHAPTER 7

The Murder and the Chase

On Wednesday, January 25, in Providence, Rhode Island, residents were abuzz about a videotape that captured a police officer kicking a black man on the ground during a disturbance at a rhythm and blues concert. The response by city leaders was swift and decisive. The Providence police chief suspended the officer without pay while the apparent police brutality was investigated, and he did not mince his outrage: "I'd like to fire him if everything is as it appears to be on that tape," the chief said.

On Wednesday, in Boston, the city's mayor, Tom Menino, was trumpeting new crime statistics showing a marked drop in homicides, aggravated assaults, auto thefts, and burglaries. The eighty-five murders in 1994 represented a 13 percent decrease from the ninety-eight murders in 1993. The mayor called Boston a "safer city," although residents of Roxbury and Dorchester told reporters they still lived in fear.

Ironically, on that same day, Boston's police commissioner was finalizing disciplinary action in a tragic controversy hounding the department. The previous March a botched drug raid had resulted in the death of a retired minister. Using a no-knock warrant, heavily armed officers from the SWAT team and the drug unit barged into a fourth-floor apartment in Dorchester. The only person inside was the seventy-five-year-old Reverend Accelyne Williams. Officers chased the minister, wrestled him to the floor, and handcuffed him. The frail Williams suffered a heart attack and died. The stunning mistake made national headlines. The minister's widow sued the city for $18 million. Ending an internal investigation that fingered breakdowns in supervision, Commissioner Paul F. Evans was busy preparing for a press conference during which he would announce the suspension of a lieutenant and reprimands for two other supervisors.

Earlier that very morning, around two o'clock, after gunshots sent patrons of the Cortee's scattering into the night, either running on foot or riding in cars past onrushing teams of gang unit and other Boston police officers, Lyle Jackson and his two friends headed over to Walaikum's Burger.

To get there, Lyle's friend Stanley drove his Hyundai up Washington Street and came within a block of where Mike Cox and his family lived on Supple Road. Stanley continued on past a high school, past Castlegate Street, and past a fire station. Washington Street then emptied into the Grove Hall section of Roxbury. The three turned right onto Blue Hill Avenue, passing the red-brick Muhammad's Mosque #11, the Boston headquarters for the Nation of Islam founded in 1954 by Malcolm X.

The drive was no more than a half-mile long. Walaikum's was now practically in front of them, just a block up Blue Hill Avenue—451 Blue Hill Avenue—across the street from a hairdresser's, a fashion store, and a used-furniture shop. Walaikum's was a hole-in-the-wall with a shabby storefront. On each side of the entrance, two air conditioners stuck out of the front wall. The neon sign hanging above the store was hokey-looking—three palm trees swayed into the squiggly letters spelling "Walaikum." The door opened into a small room. Directly ahead was a counter where food orders were placed amid the racks of bread. To the left, four tiny tables and a bench were squeezed together, and, to the right, another counter where patrons could stand and eat. The only telephone was in the kitchen.

Lyle and his friends found the restaurant filled with other young people who'd piled out of the Cortee's. Lyle ordered chicken wings and a hamburger. He was a steady customer and recognized many of the faces in the crowd, but mostly by nicknames, like Flavor and Pooh. The atmosphere was charged and noisy. In the kitchen, owner Willie Wiggins worked furiously to turn out the orders. In the crowd, Lyle's hamburger accidentally got knocked to the floor. He went to the counter to order another. Everyone was still talking about the gunfire at the Cortee's. "You can't go out anymore," joked someone near Lyle. "It's getting like New York around here."

While many clubgoers resurfaced at Walaikum's, the contingent of Boston police hung around outside the Cortee's, the sizzle gone from their night. The flat mood was apparent in the conversation on the police radio channel. "Okay," Craig Jones muttered in a monotone, "what's the game plan?" The gang unit supervisor, Sergeant Ike Thomas, replied, "The game plan? They took our ball and bat." There was radio silence—the equivalent of no comment. "Everybody into the base," commanded Thomas.

The gang unit and other police officers began clearing the scene.

Dave Williams, meanwhile, made his way back to the station in Dorchester, where he ran into Jimmy Burgio in the parking lot, pumping gas into his cruiser. They got to talking. Williams mentioned he was thinking about sitting on a house off Bowdoin Street, not far from the Cortee's, known for late night parties and trouble. "See what was going on," Williams said. Burgio was interested and agreed to ride along with him though they'd never teamed up before. Burgio left his cruiser in the lot and climbed into Williams's. They didn't bother telling their sergeant about their plan. Before driving back across the district, they swung by a Dunkin' Donuts for two coffees to go.

Elsewhere, Richie Walker was no longer torn about being tied up with a speeding car arrest and not ever making it to the Cortee's. He overheard the sour turn of events. Walker now sat with the

Peugeot awaiting a tow truck so he could clear the scene, return to the station in Mattapan, and write up the paperwork on his arrest.

In yet another part of the city, Kenny Conley and his partner, Bobby Dwan, operated on a different radio channel and hadn't even heard about the failed mission at the nightclub. But they were coming off their own small setback—the fruitless hour-long stakeout of a building to catch a supposed drug shipment to apartment 3. The tip a worried tenant had given them proved to be nothing. It was 2:07 A.M. when they took off.

Within a minute of leaving, Kenny and Bobby overheard that another patrol car had stopped "suspicious persons" in the parking lot of a liquor store several blocks away. The radio chatter included mention of drugs and prostitutes. It was no big deal, really, but Kenny turned the cruiser in the direction of Blanchard's. "We like to outnumber them," Bobby said, "so we swung over."

By 2:30 A.M., the dozen or so members of the gang unit had filed into their office in a nondescript building on Warren Street. "Lick our wounds, so to speak," said Mike Cox. The unit was not accustomed to coming up empty. "We were usually pretty successful, you know, working a Friday or Saturday night and arresting two or three people for doing things," he said. "To work with that many of us together and not get anybody, and when you hear shots fired—it was more than frustrating."

The office was located upstairs in the two-story brick building with tinted glass. Warren Street ran north from Blue Hill Avenue in Grove Hall to Dudley Square, where the Roxbury police station known as B–2 was located. The building was also about equidistant between Walaikum's and Winthrop Street where Mike grew up.

Mike and the others talked a bit about what went wrong, but mainly began to "break down" for the night so they could head home. Some of the guys sat at desks, others were in the locker room or the bathroom cleaning up. They filed some paperwork. Before leaving, each would leave behind the keys to their unmarked cruisers.

It was pretty quiet, except for occasional crackle and buzz from the handheld radios they usually kept clipped to their belts. Most were turned to channel 3 because that was the channel covering Roxbury and Dorchester, the busiest areas in the city, crime-wise. The office had the feel of a losing team's locker room. There was no way around it: Hip-Hop Night had been a bust.

"What a waste of time," Mike said.

This much they knew.

What they didn't know was everything was about to change.

Smut and Boogie-Down were already standing at the bar when Smut heard Tiny and Marquis outside Conway's looking to be let in. Smut went to the door. He was thinking the shooter at the Cortee's must have been Little Greg even though he had not gotten a look inside the car. "Tiny had no beef with anyone else." Smut opened the door and Tiny rushed inside all hyped up.

I told you, I told you, Tiny said. Talking fast, he told Smut and the others he'd seen Little Greg in the front seat, next to the driver. I told you he was up to something, he said.

They all had a drink. Smut was feeling drunk—not staggering drunk, but a mellow, feel-good buzz. He wanted things to settle down. Most of all, he wanted to call it a night and get home to Indira. But his interests had to be melded with the crew he was with—that's just how it worked—and a roundabout conversation ensued as they debated a plan that would respect everyone's needs. In the end, they settled on a plan that seemed logical to them—at least for that hour of the night.

Tiny felt he needed to get Marquis home. But that meant driving all the way back up Blue Hill Avenue to get to where Marquis was staying near Dudley Square. Smut felt a similar obligation to Boogie-Down. "He'd been riding with me all night," Smut said. "He was my responsibility." Boogie-Down's destination was close to where they'd just been; his girlfriend's apartment was walking distance from the Cortee's.

But the last thing Smut wanted to do was to head back up Blue Hill Avenue, only to turn around and drive back down to get to Indira's. It was all so circular—so Smut had an idea. He suggested that since Tiny was going to be making a big loop to drive his younger brother home, he could take Boogie-Down too.

Smut and Tiny got into a little argument. Tiny didn't like the idea. He chided Smut, saying Smut was always trying to get out of giving rides home. He also played the birthday "card," reminding Smut it was his birthday. Tiny said if Smut wanted him to give Boogie-Down a ride home, Smut should ride along. You roll with me, he said.

The guilt trip worked. The foursome left Conway's. Smut followed Tiny to Tiny's mother's house, where he left his car and jumped into the back of the gold Lexus. Tiny took advantage of the pit stop to run inside to retrieve a silver 9mm Ruger semiautomatic pistol he had hidden there. With the added protection, they took off. Smut was riding in the backseat behind Tiny, and Boogie-Down was behind Marquis. There were now two handguns in the car: Tiny's silver Ruger and Boogie-Down's.

The ride actually went quickly. Few cars were on Blue Hill Avenue at two o'clock in the morning. Smut was slumped in the backseat, tired and boozy, and he slumped even further when Marquis announced he was hungry and wanted to go to New York Pizza. Smut was not interested in making any stops besides the mandatory dropoffs. He was glad when Tiny vetoed the pizza joint: too far out of the way. But then someone said Walaikum's was up ahead and that it would be open.

Walaikum's was also in the heart of that night's darkness—near the Cortee's, near Castlegate Street, and near where Little Greg lived around the corner.

"Dudes wasn't thinkin'," Smut said later.

Tiny slowed as he drove past the restaurant, made a U-turn, and parked a couple of car lengths from the entrance. The four climbed out of the car—the menacing-looking Boogie-Down in brown boots and hat, hunched in his tan jacket; the beefy Marquis, the tallest, in his black hoodie sweatshirt; Tiny with his braids and gray sweatshirt; and Smut, the shortest, in a brown leather jacket. They saw the restaurant was crowded—standing room only. When they got to the door, Smut muttered to himself, "Man, fuck this." Boogie-Down went ahead, Tiny and Marquis near him. Smut turned and figured he would just wait for the others to get some food.

Smut then noticed Tiny tensing up. Someone standing by the entrance had casually asked, "Waz up, Tiny, waz up, Marquis?" Tiny flipped out on the kid. What the fuck, saying my name out loud like that. What the fuck you thinking?

Tiny was hissing at the kid in a hushed, angry tone. Smut didn't understand why. The only thing he could think was Tiny was acting tough in front of Marquis. Smut stepped back toward the Lexus. Tiny began stuttering, and Smut saw his eyes darting between inside the restaurant and out. That's him, Tiny was saying excitedly. That's him. Tiny was indicating a husky black guy standing at the counter at the front of the line. That's him—Little Greg's driver, he was saying. Marquis pulled out one of the guns. Boogie-Down was already farther inside, heading toward the counter.

Several teenage girls by the door later said the first sound they heard was the ratcheting of a handgun's slide, as if it was jamming. The clicking sound made the girls jump off their stools, and their sudden, spastic movements rippled through the crowd. "He's packing! He's packing!" someone was shouting. Walaikum's erupted in pandemonium. Patrons were diving onto the floor looking for cover behind the few tables and chairs. Those by the entrance rushed to escape outside.

The target was Lyle Jackson. Tiny had seen Lyle and thought Lyle was Little Greg's driver. But Tiny was mistaken, and Lyle, at the counter waiting for his food, turned to see two guns aimed at him. He scrambled to get away, knocking chairs aside and stumbling to the ground.

Some said they heard two or three gunshots. Others said it was five or six.

The first call to 911 was made at 2:39 A.M. and fifty seconds. "Can I get the policemen quick," said an unidentified male.

"What's the address?"

"451 Blue Hill Avenue. There's been a shooting up here. Hurry up."

Right away the call was broadcast over the Boston police channel 3. Hearing it, Mike Cox, Craig Jones, and the others in the gang unit froze in their tracks. The radios on their belts had exploded in a chorus with the initial shooting report. "The room goes quiet when there's a call like that," said gang unit supervisor Sergeant Ike Thomas. The location was just down from their office on Warren Avenue. No one in the unit spoke, said Thomas, as they all strained "to hear what else is going to happen."

In Walaikum's, Lyle Jackson lay on the floor on his back, a pool of blood widening around him. He had three bullet holes in his chest, one on the left side and two on the right. His eyes were open, but he was disoriented. His skin paled and quickly turned cold. One girl put her jacket over him. His friend Marcello was by his side, urging him to hold on. "He was trying to talk to me. But I told him, 'Don't talk. Just fight it. Stay a little.'" Stanley ran to get Lyle's mother, who lived around the corner on Warren Avenue.

The second 911 call was made from a nearby pay phone one minute after the first—at 2:40 A.M. and fifty seconds. This caller was a security guard who'd worked at the Cortee's and knew a trick or two about jacking up the police response.

"I've got an officer down in Walaikum's," he said, "Walaikum's on Blue Hill Avenue."

The lie was tantamount to yelling fire in a crowded theater. In quick succession, a series of urgent calls went out over the police channel 3. "We have an officer down," the dispatcher said. Then the dispatcher said, "451 Blue Hill Avenue. Officer down," and a few seconds later repeated, "Officer down," but added: "There were shots fired."

In an instant, members of the gang unit were on their feet and heading down the stairs. "Everybody just ran out the door," said Gary Ryan. Ryan jumped behind the wheel of one unmarked cruiser while his partner, Joe Teahan, climbed into the passenger side. Don

Caisey got behind the wheel of a car carrying Sergeant Thomas and another officer. Mike and Craig jumped into their cruiser. Craig was in the driver's seat. It was a moment when many different thoughts raced through Mike's mind. "I hope he's not hurt bad. I hope he's not shot. I hope I don't know him. I hope, you know, it's a mistake."

They were not the only ones responding. Police officers everywhere were on their way to Walaikum's. The reason was a mixture of human nature and the solidarity of the cop world—a call about one of their own in trouble, said Mike, "would bring out more police officers than would normally come."

Dave Williams and Jimmy Burgio, coffees in hand, were approaching the party house they'd decided to stake out when they heard the call. They got on the radio with the dispatcher, identifying their car and saying, "We're heading up." Initially there was confusion about the restaurant's name, with the dispatcher calling it the M & M Tavern, which was also on Blue Hill Avenue. Williams and Burgio overheard another officer jump on the radio and straighten the dispatcher out, saying the tavern was "all closed up." In short order, the dispatcher had the correct name. The key piece of information was the address: 451 Blue Hill Avenue. Burgio had never been to Walaikum's before, but Williams knew where to go. "Everybody's coming, you know, there's a police officer shot," Williams said.

One of the officers who set off for Walaikum's was Ian Daley. Daley, in his sixth year on the force, was born in England and moved to Boston when he was a toddler. He joined the force after graduating from college and had worked mostly in Roxbury in a one-man service car—the Bravo 431. He was at the police station in Dudley Square finishing writing a report when he heard about the shooting. Daley immediately ran outside and got into his cruiser.

The call was now going out on every police channel, not just channel 3. Kenny Conley and Bobby Dwan had just pulled out of the liquor store's parking lot, done with serving as backup in the handling of the "suspicious persons." They looked at each other. "You never know who it is," Bobby said. "Could be your brother, could be your friend." Kenny tried to get a fix on the shooting's location. Grove Hall was south from where they were—on the other side of Dudley Square. It was close by, but neither he nor Bobby was familiar with that area. Even so, Kenny got on the radio to report the Delta K-1 was in the area and "going in." He activated the car's siren and blue lights. "We took off," Bobby said, "adrenaline pumped, you know, we're flying." The two hoped that by monitoring the dispatcher's play-by-play, they'd be able to get their bearings and pitch in.

Richie Walker, listening from his cruiser as he awaited a tow truck, knew exactly where Walaikum's was located. But he decided against racing immediately toward the shooting scene. He adopted a wait-and-see strategy of monitoring the radio for any developments—particularly any news about the direction of the suspects' escape.

Given that Walaikum's was in their district, officers from Roxbury were the first to arrive. Jimmy Rattigan and Mark Freire, partners in the Bravo 101 car, were only several blocks away dealing with a stolen car that had been torched. "It was still burning when we got there, and the fire department was coming," Rattigan said. The two were known on the street as Rocky and Bullwinkle. Rattigan was taller—topping six feet—and he weighed about 270 pounds. Freire was at most five-ten and weighed 190 pounds. They'd always worked in Roxbury and became a team shortly after Rattigan joined the force six years before. "We kinda hit it off, kinda policed the same style," Rattigan said. By that he meant they were pro-active. "We'd climb trees, we'd climb rooftops, we'd hide in bushes to catch these guys.

"Everyone knows cops who won't get out of the car in pouring rain—and me and Mark would get out and just start walking the hallways of the projects and catch somebody with a pistol and drugs and stuff. So we were always aggressive. Something like this—a police officer shot—we gotta go give it 100 percent."

Rattigan and Freire pulled up to Walaikum's behind another

officer from their station—Ronnie Curtis, who'd also been at the car fire. Curtis ran into the restaurant first, dodging bystanders. Rattigan looked and could see the victim. "You could see his legs inside the front door, pointing out towards the sidewalk." Blood was everywhere. "It was one of those things," Rattigan said, "you knew right away somebody's gonna be dead."

Curtis was yelling on his radio, "Get us an ambulance down here! Get us an ambulance down here!" The urgency in his voice confirmed for Rattigan the seriousness of the victim's condition. He and Curtis had once worked together as emergency medical technicians. "It's pretty bad if Ronnie's saying that, because he's a pretty calm guy."

Rattigan was next to his cruiser when a car driven by a security guard pulled up. The driver said he'd seen the shooters take off in a gold Lexus and that he'd gotten a partial license plate number. Rattigan was on his radio right away.

"Bravo 101," he yelled. "Bravo 101."

The dispatcher replied, "The 101. Come in, 101."

Rattigan told the dispatcher what he had, and the dispatcher was immediately broadcasting on all channels. "7—6—2," he began. The plate number was actually incorrect, but that was what the guard had thought he'd seen. The car's model, however, was on the money: "Gold Lexus. Heading down Warren."

Other officers began pulling up. Rattigan yelled for Freire over by the restaurant's entrance. "We still believed it was an off-duty police officer shot because neither one of us made it in to see if we recognized him." The two climbed into their cruiser and sped off to search for the fleeing gold Lexus.

When Dave Williams and Jimmy Burgio arrived, they found the scene chaotic and bystanders screaming, "He got shot, he got shot." Burgio saw the bloodied man on the ground and thought he recognized him. He would have bet money it was Craig Jones. He yelled this to Williams, but Williams looked and knew right away this was wrong; the victim was not Craig. Williams also knew the victim was not a cop. He called in the update, that the shooting victim at Walaikum's was not an officer down.

The officers arriving at Walaikum's faced a number of key tasks at hand—attend to the shooting victim, secure the crime scene, gather witness information, find out where the shooters went. Williams and Burgio did some of each, talking quickly to the owner and then a few bystanders. The police channels were crackling with calls to be on the lookout for a gold Lexus, and when Williams and Burgio caught wind a pursuit was in the making, they wanted in. They hopped into their cruiser, leaving the crime scene to the officers from the Roxbury district.

Mike Cox and Craig Jones had just pulled up to Walaikum's, no more than a minute or so after they'd left the gang unit, when they heard the update about the victim. "So we knew right away there wasn't a policeman shot," Mike said. But they hadn't yet heard about the gold Lexus. They spotted other police cruisers and an ambulance. Craig headed toward the restaurant's entrance. Mike stayed outside. "We were trying to find out what kind of car the shooters were in."

Paramedics worked on Lyle Jackson, noting, in a report prepared later, Lyle was suffering from "multiple gunshot wounds," "a very serious loss of blood," and was "in a great deal of pain." They immobilized his neck and back, gave him oxygen, propped up his legs to get blood to his brain. The diagnosis was "acute, major trauma."

Lyle's mother, Mama Janet, arrived as they were putting Lyle into the ambulance. She'd run from her house with another son. She saw Lyle and called out his name. "He kind of looked at me," she said. "His eyes were closing. He looked at me and the tears started rolling down his cheeks." Lyle Jackson lasted six days before dying on January 31 at Boston City Hospital.

When he'd heard gunfire, Smut bolted to attention—suddenly feeling cold sober. He watched as people began screaming and running from Walaikum's. The other three were back, "huffing and puffing." He yelled at Tiny to get going. "Pull off, pull off," he ordered. Marquis was in the front seat, and Boogie-Down was next to Smut in back. They drove a block down Blue Hill Avenue and turned right onto Warren Avenue. Within a few blocks they turned left onto a side street, avoiding any oncoming police cars.

Smut and the others yelled and swore at one another for taking the Little Greg dispute too far and shooting up Walaikum's. They also quickly decided it would be best to split up, and the first idea was to get Marquis home because he lived close by. Tiny began winding away from Grove Hall toward Dudley Square on what Smut considered "back roads." They worked their way through a thicket of streets either intersecting or near Humboldt Avenue, the Roxbury boulevard known as the location of one of the city's most notorious murders, the 1988 shooting death of a twelve-year-old girl named Tiffany Moore. Humboldt and its side streets were on the north side of Franklin Park, originally the "crown jewel" of Frederick Law Olmsted's network of parks created throughout the city a century before. The 527 acres were now in the middle of the city's poorest section and, while featuring a golf course and a zoo, always seemed in need of an overhaul.

Being inside the Lexus was like being inside a bubble. Smut, Tiny, Marquis, and Boogie-Down had no idea of the size and scope of the police response to the initial report that an off-duty cop had been shot at Walaikum's. They had no idea that throughout the city nearly every officer on duty was listening closely to the radio while those in the immediate area were either racing to Walaikum's or looking for them. Ian Daley was among the latter. He'd left his paperwork behind at the station and, instead of racing to Walaikum's, began cruising the outskirts trying to think where the shooters would go. He wasn't alone—Dave Williams and Jimmy Burgio, Gary Ryan and Joe Teahan, and Jimmy Rattigan and Mark Freire were all driving around Roxbury looking for the Lexus.

Smut, Tiny, Marquis, and Boogie-Down did not know about any of this. Nor did they realize that when they made a turn onto Martin Luther King Boulevard they were spotted—almost simultaneously—by two security guards riding in their company car and by a Boston police officer in his patrol car. The security guards immediately began broadcasting their location over a radio frequency used solely by their company.

Mike Cox was standing outside Walaikum's when a black man dressed in a dark uniform approached him. He was a security guard named Charles Bullard. Bullard excitedly began telling Mike that two guards from his company were following the car—the gold Lexus. Mike asked what he meant. "They're chasing the car?"

Bullard held up his radio and Mike listened to the voices talking about the Lexus. None of the police channels had yet broadcast information on the Lexus's whereabouts. Mike summoned Craig. They told Bullard to get into the backseat of their cruiser. Craig jumped in behind the wheel and Mike was on his radio.

"TK," he said.

"Okay, come in," the dispatcher said.

"One of those security officers is in the car with us," Mike said. "They seem to be chasing the car that did this shooting. He's got the radio with him and we're listening to this chase, trying to catch up to it."

Mike and Craig were at Walaikum's for all of two minutes. They headed down Warren Avenue hoping to get a bead on the Lexus.

The first Boston police officer to see the Lexus was an officer from the Roxbury station named Dave McBride. McBride was driving down Martin Luther King Boulevard when he heard Jimmy Rattigan putting out a description of the "suspect vehicle," and there it was—the gold Lexus. The Lexus wasn't speeding, just motoring down the street. Behind the Lexus, McBride saw another car, the one with the two security guards in it. McBride did not activate his lights or siren, but he called in his location.

The first sighting by the Boston police brought a sudden focus to the manhunt. For some officers, it meant realizing they were nearby. For others, it meant they were way off track, like Kenny Conley and Bobby Dwan. They'd raced into Dudley Square but then got all mixed up. Confused, they made turns that took them farther away from the action. It didn't help when they heard Dave McBride start "calling off" the Lexus's whereabouts. "We didn't know the streets," Kenny said. They couldn't make any sense of the information and took turns yelling out in frustration: "Where the fuck are they!"

Mike and Craig were not lost, but the moment they heard McBride reporting the Lexus's location, they knew they were off the mark. They'd gone up Warren Avenue toward the gang unit and Dudley Square, but the Lexus had now reversed direction and was working its way back south toward Franklin Park. With McBride reporting the Lexus's movement, Mike turned to Bullard and told him to turn off his radio. "His radio was still blaring and our radio has two different signals, and it was too confusing." Craig turned around so they could work their way toward Franklin Park.

It wasn't until Tiny approached a traffic light that he discovered the Boston police cruiser in the rearview mirror. He hadn't noticed the cruiser fall in behind him; it seemed to have appeared out of nowhere. Tiny turned off his headlights, and he didn't stop at the light.

McBride watched the Lexus go dark and accelerate. It was time: He turned on his siren and blue lights. What had been a watch-and-wait shifted abruptly into an actual police pursuit. McBride was on his radio shouting out his location.

"We got a Lexus going down—uh—Crawford!"

Jimmy Rattigan and Mark Freire, listening to the frantic tone in McBride's voice, realized the Lexus was heading their way. They were on Humboldt Avenue. Crawford Street intersected Humboldt. The two were thinking they could cut off the Lexus. "We still thought a police officer was shot," Rattigan said.

Rattigan turned onto Crawford. He ignored the fact Crawford was a one-way street lined with parked cars. But he was forced to pay his full attention to what he saw coming his way, "This car just flying, I mean, it was moving." They had not expected to see the Lexus so quickly. "I'm like, holy shit!" Rattigan had a split second to make up his mind. "I'm either gonna let this guy hit me head-on, and they were going a lot faster than we were, or I gotta get out of the way."

Rattigan saw there was an opening between two parked vehicles. It wasn't a full space, but maybe big enough to make it in. He turned sharply to the right to avoid the head-on collision, but the space was too small and he crashed the cruiser into an empty van. The two cops were thrown around the front seat of their car as the air bag exploded. "The cruiser was destroyed," Rattigan said. He wrenched his neck, twisted his back, and suffered minor burns on his forearms from the air bag, but he didn't feel anything at first.

Instead, Rattigan leaned hard into his door trying to get it to open. It had buckled and wouldn't budge. The Lexus slowed to navigate past the crash. Rattigan looked over at the driver, who was not much more than an arm's length away. "We locked eyes," Rattigan said. "I'll never forget looking at that face. He didn't care."

While Rattigan was stuck in the car, Freire squirmed out of the passenger's side. He pulled a 9mm semiautomatic Glock handgun from his holster. "Mark was on him with his gun," Rattigan said. Freire aimed at the moving car. Rattigan could practically feel his partner's struggle over whether to shoot at Tiny Evans. The driver was behind the wheel of a getaway car. But Freire did not have enough to go on at that moment: who the driver was and his role, if any, in the shooting. Freire began grunting and growling, as if wrestling with his weapon. "Literally in seconds a thousand thoughts go through you mind," Rattigan said. "What do you do?"

And in those seconds, the Lexus rolled past them and began picking up speed.

Rattigan pushed his door open enough to be able to see the Lexus's taillights. Then he watched several Boston police cars go by while the security car stopped to help. Freire was standing over on the sidewalk. "He's screaming, 'Fuck,'" said Rattigan, "and I'm like, 'Fuck!'"

Rattigan slumped in his seat. He and his partner were out of the chase, done for the night. Adding insult to injury, Rattigan watched a resident come storming out of one of the apartment houses to complain about his van. The man even got a camera to photograph the damage.

He never asked about the officers' injuries.

Making its way past the crashed cruiser, the Lexus picked up speed as it continued down Crawford and turned right onto Elm Hill Avenue. Four blocks later, Tiny turned left onto Seaver Street, a wide, two-lane corridor bordering the northern side of Franklin Park.

The police chase was moving up a notch. While most officers by now knew a civilian, not an off-duty cop, was the shooting victim, they'd all just heard on channel 3 the frantic yelling, the radio static, the sudden radio silence, and then Mark Freire calling for an ambulance for him and his partner. The getaway car meant business.

"The suspects, these are murder suspects," Mike Cox said. "They're on the run. They're obviously scared. One cruiser down, and we're trying to figure out where the hell they're going."

Tiny had exploited the crash to put a few blocks between his car and the nearest police vehicle. Indeed, McBride had fallen off the chase, as had the security company car. But Ian Daley was on Seaver Street driving alongside Franklin Park when he spotted the Lexus as it came down Elm Hill. Daley watched the Lexus turn onto Seaver Street and saw that it was alone.

Daley radioed in the Lexus's location. He was the first to call in the car's complete, correct license plate: 676 ZPP. Soon enough, police cars from all over were looking for ways to join what was developing into the largest and longest-lasting police chase anyone in the department remembered. Jimmy Burgio, for one, later said "only on TV" had he ever seen anything like it. Mostly Boston police officers participated, but state troopers and officers from the city's housing police and the municipal police, nicknamed "munies," also got involved.

The high-speed chase is one of the ultimate cop moments, carrying the hugest of rushes. But such chases also frequently get the better of the participants. Or, as one police expert dryly put it: Officers involved in a pursuit too often "do not conduct themselves consistent with their training nor written directives." This chase was no different. Boston police rules requiring supervision of high-speed chases, particularly at their conclusion, were words on paper signifying little. It was Wild West time in Roxbury.

Ian Daley turned on his siren and lights. He saw the cars' occupants looking around. "You know, one head would pop up, another head would pop up, and one would go down." He pulled up behind the Lexus.

Daley wasn't the only one who saw the gold-colored car. Dave Williams and Jimmy Burgio were approaching from the opposite direction. They'd made the right onto Seaver from Blue Hill Avenue and spotted the Lexus. Williams saw the Lexus's lights were turned off and that Daley's cruiser was behind it.

Williams thought he might be able to cut the Lexus off if he jumped the median. But he quickly decided not to—the concrete strip was eight inches or higher, and he'd more likely get hung up on it than make it over.

Instead, Williams shifted the cruiser into reverse and roared backward. Looking over his shoulder, he saw other cruisers, including a munie car. Williams had to slow down to navigate his way back to Blue Hill Avenue. The slow-down incensed the intensely competitive Burgio. When they almost collided with the munie car, he screamed out the window at two municipal officers, "Get out of *my* chase!"

Tiny turned off Seaver Street onto Blue Hill Avenue, heading in the direction of Mattapan. Ian Daley followed, making the right turn onto Blue Hill. Williams, seeing an opening, jumped in behind Daley. Other cruisers fell in behind the two lead cars. Burgio saw cruisers behind him "as far as I could see."

Daley radioed in the new location—Blue Hill Avenue—a street so well-known it was like saying Broadway to a New Yorker. When they heard it, for example, Kenny Conley and Bobby Dwan erupted with the click of recognition. "Now we know where to go," Kenny said. They'd been monitoring the chase feeling useless, but Kenny was now flooring the accelerator and hitting speeds of 90 mph as they raced up Columbia Road, which emptied onto Blue Hill Avenue at Franklin Park.

Richie Walker, meanwhile, listened to the progress of the chase and began thinking if it continued heading south in his direction, this might be something he'd want to join. Mike Cox and Craig Jones, for their part, turned onto Blue Hill knowing they were getting back on the right track. They saw no sign of the chase ahead of them, but headed down Blue Hill, playing catch-up.

Tiny drove down Blue Hill along the east side of Franklin Park, past the entrance to the zoo, where construction was scheduled to begin soon on a new exhibit to house a dozen African lions, marking the return of lions to the zoo after a twenty-five-year absence. Past the zoo at the first intersection, Tiny turned right onto American Legion Highway. The others bounced around inside. They were all rattled by the growing line of cruisers behind them. For his part, Smut decided he was going to have to direct their escape. Turning down American Legion Highway, he thought, was a bad move. Since it was straight and wide open, he figured no way they'd be able to outrun the police on it. What they needed to do was to get into the side streets.

Then they saw trouble—flashing lights up ahead, where police were hastily setting up in a roadblock. Tiny braked. Smut was yelling at him to turn around. Tiny crossed the grassy median dividing American Legion Highway and began driving back toward Blue Hill Avenue. Behind them, Ian Daley copied Tiny's moves.

Farther back, Dave Williams watched the two cars slow down to execute the maneuver. He thought, "Okay, yeah, we got him." He began driving across the grassy median too and told Burgio to get ready. "We're going to ram this car." They snapped on their seat belts. Burgio opened his window to eliminate the possibility of shattered glass flying all over them. They braced for a collision.

But then the Lexus was gone. It had taken a right down a side street, Franklin Hill Avenue. "I was about to do it and it was like a black hole appeared," Williams said.

Smut had ordered Tiny to turn onto a street running along the south side of the Franklin Hill housing project. It was where Smut lived as a boy, and the Lexus was soon speeding past the actual building where his family's apartment was located.

Instead of the Lexus, Williams was bearing down on Ian Daley. The two cruisers skidded to avoid colliding. Daley turned right and Williams followed. Daley kept yelling out locations on the radio, "He went over the median here! He went over the median here!" Then, seconds later, he yelled, "Franklin Park project! Franklin Park project!"

The dispatcher called out the update: "Franklin Park project now. Looking for a gold Lexus. 676 ZPP."

Mike Cox and Craig Jones were making their way down Blue Hill Avenue trying to get their bearings. "We weren't really going fast," said Craig, "because we were listening to the transmissions."

When Mike heard that the Lexus had turned down American Legion Highway, he had a hunch. "They're probably going to Franklin Hill," he told Craig.

Mike was putting himself in Smut's mindset. He wanted to anticipate the Lexus's next move. It was a game, of sorts, where the checkerboard was the streets of Roxbury. He and Craig did not want to follow the others and turn down American Legion Highway; instead, they wanted to make a calculated guess where the Lexus was going.

With that in mind, they picked up their pace. They turned off Blue Hill Avenue onto Harvard Street. The right turn took them past a red-brick building where for the past two years Boston Red Sox slugger Mo Vaughn ran a youth program for city teens who gathered after school to work on their homework and eat a catered meal.

They turned knowing that Franklin Hill Avenue, once it made its way through Smut's boyhood housing project, intersected Harvard Street. Mike was even thinking the shooting suspects might bail out. Projects were often where suspects looked to shake cops pursuing them.

Instead, within seconds of making the turn onto Harvard Street, Mike saw the gold Lexus. It was their first sighting of the night. The car was flying down Franklin Hill Avenue. "Literally kind of come off the ground and come down the hill," Mike said. He didn't see any police cars behind it, but he could see lights from the cruisers reflecting in the night sky.

Mike and Craig, and the Lexus were perpendicular to each other—Mike and Craig on Harvard, and the Lexus coming fast down Franklin Hill. "It came directly in—virtually right at us at the intersection," Mike said. The Lexus roared into the intersection, skidding onto Harvard.

Suddenly, said Mike, "we are side-by-side." He saw four black men inside the car. He and the driver exchanged looks, and Mike noticed the driver wore his hair in braids. The two cars drove parallel for a few seconds, and then the Lexus cut left.

"The car tried to ram us," Mike said.

Craig swerved into the oncoming lane. "It just barely missed us," Mike said. Craig had another worry—the weapons. "I was basically thinking that, you know, I don't want him to start shooting into our car." He jammed the brakes to slow the cruiser. The Lexus sped ahead of them. Craig then turned the steering wheel to the right, and the cruiser moved in behind the Lexus. The two cars were speeding down Harvard Street.

"Okay, we're the lead car," Craig yelled on the radio.

Dave Williams and Jimmy Burgio did not seem happy about falling from second place behind the Lexus to third—as if someone cut in front of them in the lunch line at the school cafeteria. "The unmarked cruiser cuts right in," Williams noted later. Burgio was peeved; the way he heard Craig's broadcast was: "The *real* cops are in the lead now." The mistaken version played into his view of Craig as glory hound.

It was true, though. Mike and Craig had indeed become the first police car.

CHAPTER 8

The Dead End

The lineup at the very front was now established, and it stayed that way for the remainder of the chase. Behind the Lexus were Mike and Craig, then Ian Daley, and then Dave Williams and Jimmy Burgio. In a pursuit, the lead police car customarily radioed in the route, and, for a few seconds, Mike and Ian Daley "stepped on each other," or talked at the same time. The radio transmissions were briefly confusing and clogged, until Daley went quiet and ceded the floor to Mike.

Behind the first three Boston police cars, the lineup changed as new cars joined the lengthening conga line and others fell off. Richie Walker, for example, did decide to abandon the Peugeot and head down toward Mattapan once he heard Mike Cox describing a winding route into that neighborhood. Eventually, Walker jumped in line behind Williams and Burgio, becoming the fourth police car. Likewise, Gary Ryan and Joe Teahan of the gang unit, who'd not been directly involved for much of the chase, eventually ended up among the first half-dozen police cars behind the gold Lexus.

Kenny Conley and Bobby Dwan, meanwhile, were racing down Blue Hill Avenue, playing a hunch the Lexus was headed to Mattapan Square. Having grown up in Mattapan, Bobby directed Kenny to stick to Blue Hill as the fastest route, a nearly three-mile ride

that took them up the avenue's hills with its faraway views of the
Blue Hills in the town of Milton south of Boston, and then along
flat, low-lying stretches past storefronts, hair salons, churches, and
cash-checking stores. Nearing Mattapan Square, they passed Sim-
co's on the Bridge, the famous hot dog stand from the 1930s featuring the "World's Largest Hot Dogs," which was a favorite of
Smut Brown's; his family's West Selden Street home was only a few
blocks away.

It was a chase whose speed ebbed and flowed. "On major
streets," said Mike, "certainly it was a high-speed chase. On small
streets, it slowed down quite a bit." Burgio noted the oddity of
the Lexus sometimes using its directional signal prior to making
a turn. "Probably the most courteous kid I was ever behind," he
said. Mike and Craig had a blue light flashing on the dash, but it
meant nothing to the men in the Lexus. In fact, several more times
police cruisers came at the Lexus head-on, and each time the cruisers pulled aside to let the Lexus go.

The number of police vehicles kept increasing—ranging from
twenty to forty, depending on the officer talking—as most cops
still thought a fellow officer had been shot. Many spoke later with
amazement at the number of cruisers. During one stretch, Dave
Williams marveled at the scene: "You could look back in your
mirror and all you can see is just a sea of blue.

"I said, 'Jimmy, damn, look at that. You've never seen that
before, you know what I mean?' It was like, you know, all you could
see, as far as you could see looking back, was just blue lights, and we
were just—just glowing."

Technically speaking, Mike and Craig, operating an unmarked
cruiser, were not supposed to be the front car. Section seven of the
police department's "Rule 301: Pursuit Driving" said: "Department
policy shall be that marked units lead a pursuit, wherever practical. Therefore, unmarked units involved in a pursuit shall yield to a
marked unit." The key word in the department's regulation seemed
to be "practical." Craig later said they became the lead car along

Harvard Street "not by choice," and that yielding would have been impractical. Their supervisor, Sergeant Ike Thomas, rejected as outrageous the notion that Mike and Craig should have pulled over or that he should have commanded them to do so. "In the middle of a twenty-minute pursuit it would have been extremely foolish on my part to interrupt a car chase that was involved in a very serious crime, to interrupt and say switch up and let a marked take the lead," he said.

But as the chase continued toward Mattapan Square, there was some jockeying in front. Twice Craig had to call off another cruiser that was either getting too close or trying to take over the lead. "Okay, we're the lead car; you better get away from us," Craig said, taking the radio from Mike to call out the warning.

Minutes later, he was on the radio again: "Get behind me. Don't hit me, please!"

In the Lexus, Smut was now the backseat driver, commanding Tiny to take this turn or that. To get off Harvard Street, he ordered Tiny to turn right, then right again and then another right, a zigzagging southerly route taking them deeper into Mattapan. This was Smut's turf, and he had begun to formulate a getaway plan in his mind. He knew about a dead-end street named Woodruff Way that bordered the east side of the former Boston Sanatorium, which most called the old Boston State Hospital. Smut knew the area well because West Selden Street bordered the opposite side of the wooded fifty-one acres of city land. He'd grown up riding bikes and cutting through the grounds. Woodruff Way was also one of a handful of streets that made up a housing project known as Morton Village and, as a dead end, was used by car thieves to dump stolen cars.

To get there, Smut needed to direct Tiny to the other side of Mattapan Square. But first there was a matter of the incriminating evidence in the car—the guns. While the Lexus was on Itasca Street approaching an intersection, Marquis opened his front window, cocked his arm, and tossed out one of the semiautomatic handguns. The gun bounced along the asphalt and came to rest on the lawn of a corner house.

Mike saw the projectile and was all over it. "He threw something out on Itasca!" he yelled. "He threw something out on Itasca!" Listening was a police officer named Roy Frederick, who lived a few doors down from the intersection. Frederick was off duty and up late, glued to channel 3 and listening to the amazing chase. He immediately hustled outside, looked around his neighbor's lawn, and spotted the silver-plated 9mm Ruger. He called in that he'd recovered one of the suspects' weapons and stayed put to secure the scene.

Seconds later, Marquis got rid of the second gun. It hit a minivan and then landed in the driveway of 235 Itasca Street. The Lexus had shed its weaponry, but once again, Mike called in the gun's location.

With Smut pointing the way, the Lexus kept winding its way toward Mattapan Square. During slowdowns that came with making a series of quick turns, Ian Daley thought the long line of cruisers resembled a "funeral procession."

Police cars were all over the place, trying to catch up or figure out a way to cut the Lexus off. Donald Caisey, driving the gang unit car carrying the unit's supervisor, Sergeant Thomas, never got in the conga line and was instead monitoring the radio, trying to stay close by. Then, in Mattapan, Caisey realized the Lexus was on the next street heading their way. He turned down the street, his siren blaring and lights flashing. He turned the car sideways with the idea of forcing the oncoming car to stop. "We were in the middle of the street moving back and forth."

But the Lexus kept coming. "The vehicle never slowed." In Caisey's mind was the earlier near head-on crash on Crawford Street that ended with Rattigan smashing into a parked van. The Lexus kept coming, "so I immediately pulled out of the way." The Lexus and the line of police cars following flew by.

Like Caisey, Gary Ryan and Joe Teahan kept working the perimeter, trying to outsmart the Lexus and anticipate its route but without much success.

It was in this stretch of the Lexus cutting back and forth toward Mattapan Square that Kenny Conley and Bobby Dwan joined the hunt. They'd roared into Mattapan and even drove past where Bobby grew up on Violante Street. The two officers were still under the impression the shooting victim was another officer, having missed the correction when it was broadcast. They raced up and down side streets; at one point they even found themselves in front of the Lexus, only to lose it when the Lexus turned down a side street; another time they found themselves a block away on a street parallel to the Lexus. Finally, they approached a main road and the Lexus drove past them. "Now we get into it," said Kenny. Not only were they into it, they entered toward the front of the line.

"We're heading toward Blue!" Mike called out.

"Headed toward Blue," the dispatcher relayed. "Be advised. That car is wanted for a shooting."

Smut had twisted the Lexus like a pretzel in and around the streets surrounding Mattapan Square, and the car was now back on Blue Hill Avenue. Smut's plan was for Tiny to shoot north a few blocks, cut east toward the Morton Village housing project, and barrel down to their final destination—Woodruff Way. Smut told the others they were going to a dead-end street encircled by a chain-link fence. He knew kids had cut a hole in the fence to make it easier to come and go. He wasn't sure if the hole was still there or had been repaired by city workers. But the point was to get past the fence and into the woods. "I told them to run towards the fence," Smut said.

It was the only way Smut could think of to shake the cops. Tiny and Marquis listened carefully. Marquis had no idea where they were at this point. Smut repeated the plan to help him out. Boogie-Down got it right away; he knew the Mattapan area. "Once we get there," Boogie-Down said, "we're all gonna get out of the car and try to get to the fence through an open—the fence is cut open. We was gonna make it through there, and cut through the woods and come out to the other side."

The Lexus took a series of rights—a right onto Norfolk Street, a right onto Morton Street. They drove past the intersection of Smut's West Selden Street and then took a right onto Woodmere Street. They'd entered the Morton Village housing project.

Every move the Lexus made, Mike radioed it in.

"Norfolk coming up to Morton," he yelled.

"Right onto Morton!" he said seconds later.

The dispatcher passed along the locations.

The chase was past the fifteen-minute mark and had covered about ten miles. For all the police power brought to bear, the Lexus had outrun the police and was now honing in on its exit plan. The adrenaline was rushing for those several dozen officers directly involved in the pursuit as well for the officers throughout the city listening to it. Despite the intensity and shouting, however, the transmissions at this point were breaking up.

"Where are we?" the dispatcher yelled. "Where are we?"

Mike replied, but his words were lost in the static and wailing sound of police sirens. Only fragments of sentences made it through.

"Projects," Mike yelled. "Woodmere."

Mike and Craig knew where the Lexus was headed. They knew it the moment they'd turned into the housing project. They'd worked the area and knew the layout of the streets—that a couple of streets looped, one to the left and the other to the right, and then met at the entrance to the dead end of Woodruff Way. They even knew about the hole in the fence; they'd chased car thieves who'd escaped on foot through it. Knowing all this, Mike and Craig could sense the chase was coming to a climax.

The dispatcher yelled, "Where are we?"

Only two words from Mike were audible: "Woodruff Way."

There was more static and a collision of voices.

"Just the lead car!" the dispatcher yelled.

Mike broke through. "Woodruff! Woodruff!"

Mike's voice was gone, then back to add that Woodruff Way was a dead end.

Then Mike was screaming: "Getting ready to bail!"

They were his last words.

Mike and Craig followed the Lexus making a right turn onto Woodruff Way. The road went downhill about a thousand feet to a cul-de-sac enclosed by a chain-link fence. The circular dead end was about thirty yards in diameter. Marking the end were seven steel posts in the ground, beneath a single streetlight.

The Lexus was screeching to a stop beneath the streetlight in the middle of the circle. Mike and Craig were no more than a car's length behind. Mike's heart was pounding. "You know you're about to run, and you just get prepared to do that. Gather up whatever you need to run, whatever you're gonna have, whether it's your flashlight or your radio or your firearm." He had been involved in police chases before, but nothing like this. "It seemed like an eternity." He was also feeling confident they had finally reached the "gotcha" moment. Immediately behind them were a slew of cruisers: Ian Daley in his, and Dave Williams and Jimmy Burgio behind Daley. Richie Walker had managed to get into the cul-de-sac behind Williams and Burgio. There was a bottleneck of police and emergency vehicles at the entrance to Woodruff Way, and in the next handful of cars after Walker were Joe Teahan and Gary Ryan, and Kenny Conley and Bobby Dwan.

Mike knew the numbers favored the cops. "It was such a long, long chase, and I knew there were several officers behind us, so I felt pretty good about being able to catch these people." For Mike, this was the moment—what being a cop was all about, split-second action, his life on the line and the public's safety at stake. The men in the Lexus, he said, "had shot someone—that certainly makes you want to catch them."

The four doors of the Lexus popped open even before the car came to a stop. Craig wanted to trap the driver inside, so he steered the Crown Victoria cruiser to the left side of the Lexus. Craig slammed on the brakes. Mike pushed open his door. It hit the Lexus. The two cars were that close. Tiny Evans jumped out from behind the steering wheel while the Lexus was still rolling. He ran around the front of Mike and Craig's cruiser toward one of the housing units on the left side of the dead end.

Ian Daley began braking directly behind the Lexus, while Dave Williams went to the right to complete boxing the Lexus in. But he skidded on an ice patch and lost control of the car. The cruiser scraped the two open doors on the right side of the Lexus and then smashed into a steel pylon.

Boogie-Down jumped out from the right side. He'd taken only a few steps when he was knocked down by the skidding cruiser. In front of Boogie-Down, Marquis met the same fate. Marquis jumped out, but when he put his feet on the ground, "I was hit immediately." The two, scraped and bruised, began crawling on their stomachs across the asphalt between their car and the police cruiser.

Behind them, Smut Brown had scrambled across the backseat to follow Boogie-Down out of the car when he saw the skidding cruiser hit his two friends. "They like disappeared," he said. Boogie-Down and Marquis had been in front of him and then, in a flash, they were gone. "My mind was racing so fast that I know I seen them there, then after that I didn't see them anymore." Smut was on his own. "I thought Boogie-Down was dead, to tell you the truth." He eyed the fence erected along the right side of the cul-de-sac. He hit the ground running. "I ran straight towards the fence."

Mike pushed the car door hard into the Lexus, trying to make enough room to get out. He'd seen Craig leap out of the car and race after the driver off to the left, but lost sight of them. The security guard in the backseat, Charles Bullard, also headed that way. Craig, in a matter of seconds, caught up to Tiny Evans. Tiny had stopped and raised his hands. Craig hit him in the face with his fist. "I ran up to him, I punched him. And I grabbed him by his arm, turned him around, twisted his arm behind his back."

In the other direction, Mike saw the front and back doors of the Lexus pop open. He watched the other suspects scrambling out of the passenger side. He thought he saw two of them heading toward the fence off to the right. He twisted his body, thickened by his clothing—the black sweatshirt and the three-quarter-length black parka.

Squeezing his way out, Mike ran behind the Lexus. The sound of screeching wheels caught his attention. He hesitated and looked back to see police cruisers braking to a halt. "I kind of glanced up to make sure it wasn't going to hit me." One cruiser skidded along the right side of the Lexus. Police sirens blasted and cruiser lights sliced up the night, and Mike thought the sight of the cruisers flooding into the dead end was a good thing. "There was more help coming."

As he ran behind the Lexus, he looked inside. "I could see the doors to the Lexus were wide open and I could see inside the car and see that there's no guns hanging out." The quick appraisal meant no one was looking to shoot at him, and the four suspects were all out.

Mike looked and caught sight of the suspect who bolted from the Lexus's backseat and was running over to the fence. He took off after the suspect, hard on the man's heels, barely hindered by the long down jacket and pumping his legs like a football running back.

Mike saw the man throw himself onto the fence and kick his legs up. The section of fencing was missing an iron bar, making the top unstable and barbed. The suspect was gaining traction and tumbling over the top when his brown leather jacket got snagged on one of the sharp prongs.

This was Mike's chance. "He's dangling from the top of the fence." Mike reached up and grabbed a sleeve. He held on to it tightly. He briefly looked at the suspect, but in that moment did not recognize Smut Brown. Mike tried pulling the suspect back, but the physics were against him. The suspect was already too far onto the other side. The last thing Mike wanted to do was let go—he had the suspect in his grip—but he had to.

The suspect dropped onto the hill on the opposite side of the fence and rolled. Mike took a step back. He was thinking about his next move, "whether I wanted to jump over the fence and get cut up or hurt versus trying to find another way."

The answer was to go straight ahead—up and over.

Mike stepped and reached for the fence. Then from behind, he felt the first blow, "a real sharp, painful blow."

He turned to his right to see what the hell was going on.

Kenny Conley and Bobby Dwan raced downhill toward the cul-de-sac and saw the snarl of cruisers ahead of them screeching to a halt. Kenny had slowed to get through the bottleneck at the entrance to the dead end of Woodruff Way and then accelerated again.

They were now the seventh or eighth cruiser behind the Lexus. Directly in front of them, they saw another officer from their station—Joe Horton—who was driving a one-man cruiser.

"It was very hectic," Bobby said. "The sirens were going. Lights were flashing, which if you look at them, they blind you. It was pretty dark other than the lights flashing. There were car doors open from people jumping out and running around."

Kenny, looking through the chaos, noticed the one suspect exit the rear of the Lexus and run toward the chain-link fence. He slammed the car brakes, and the cruiser began skidding to a stop at the top of the dead end. He did his best to keep an eye on the man running toward the fence.

Kenny was locked in—tunnel vision. In those split seconds, he did not pick up on the commotion Bobby was noticing farther down along the fence. He saw only the suspect on the other side of the fence—a man whose name he would later learn was Robert "Smut" Brown.

Bobby, climbing out of the passenger seat, had glimpsed three or four people over near the fence. The officers had surrounded someone. "I was just thinking they're cuffing him." In the other direction, meanwhile, Bobby saw Joe Horton run to the left after another suspect who was already on the ground. Bobby made the quick calculation. The guys at the fence had a suspect and probably didn't need him. "It seemed to me they were all set." Horton, in contrast, was running alone toward a suspect. The call wasn't close. "I worked with Horton," he said, and so he ran over to assist him.

Kenny hustled to the front of the cruiser and ran into Bobby.

Kenny was heading right, Bobby was heading left. They criss-crossed past each other and kept going. "The last time I saw him," Bobby said about Kenny, "he was at the fence ready to go over."

Kenny got over quickly. "There was no bar on top," he said. "I put my feet up and was kind of wiggling and I jumped." He dropped to his feet and stumbled. Up ahead, he saw the shadowy figure of Smut Brown leap off a little wall onto a street.

Kenny took off after Smut. Smut headed to the right, ran across the street and through a parking lot. He ran behind a building and up a hill through some woods. "He was probably forty feet in front of me," Kenny said. Smut hopped over a chain extended across a cement staircase, ran up the stairs, and then headed across another lot. Kenny followed.

They'd run the length of a couple of football fields when Kenny drew his Glock semiautomatic handgun. "Fucking stop!" he yelled.

Smut did, and he raised his arms. "You don't have to shoot me." Smut did not turn around. He yelled he was not armed. "I haven't done anything." Kenny ordered Smut to get down on the ground on his belly. Smut did, with Kenny's help. "I pushed him with my fore-arm on the back of his shoulder blades."

Smut did not resist. He was worried. He had seen plenty back at the fence, a stampede of cops beating a man he thought was Mar-quis. Smut was worried he was next—that this officer with the gun was "going to come and jump on me."

But it turned out he didn't have to worry. Kenny put his gun back in its holster, leaned over the drug dealer, and snapped on a pair of handcuffs.

When Mike turned to see what had hit him, he was hit a second time. His head exploded, and he could not see. His only thought was wondering why he did not feel more pain. "I just remember saying, like, Ouch, to myself." It was a strange question to be asking, as if his mind had left his body and taken up a position of clinical observation.

The first blow to the back of his head had rocked his brain, causing it to collide with the inside of his skull. The trauma triggered an inflammatory response of infection-fighting cells. Mike's head began swelling immediately, a bump the size of an egg.

The second blow then ripped open the right side of Mike's forehead. Blood began pouring from a laceration along his hairline that was nearly three inches long. Next Mike was pulled off the fence, and he fell toward the front of the marked police cruiser that was to the right of the Lexus. More blows followed, ferocious blows. Mike's radio fell to the ground by the front of Dave Williams's cruiser.

He was down on all fours, wobbly like a dog on its last legs. He lifted his head and saw a puzzling image. "It looked like an officer," he thought. But that was crazy, a hallucination. Mike looked again, but the initial impression would not vanish: It was a cop, a white cop. "He was standing in front of me." Mike tried to raise his head up higher to get a better look. But the only thing he saw was a boot coming flush into his face.

Now Mike felt the pain—pain in his face, his head, his shoulders, his back. The kick was followed by more blows. He curled his arms over his head for protection against the blows to "all sides of my body, from different directions."

He fought to stay conscious; he wanted to see who was doing this to him—and why? Blood ran from his nose and mouth. He was alternately conscious and semiconscious, and he'd lost any sense of time. The blows to the head happened so fast, but now everything seemed to be happening in a clouded slow motion.

"I don't know how long it took in actual time," he said.

Then, suddenly, it stopped. There was quiet, too. "I saw that there's no one, there's no one there." Mike was alone. He struggled to get back up on all fours. He crawled to the rear of the nearby cruiser. "I used it to lift myself," he said. "I was having trouble breathing and standing."

He tried to balance himself. His hands swished in the blood on the car's trunk—his blood. He was facing the end of Woodruff Way with the hole in the fence. Then he detected that someone was standing a few feet away. He heard the man saying something. But in the thick fog that had overtaken him, he could not make out the words right away.

Mike then realized the man was ordering him to submit to an arrest. Mike couldn't believe this. He looked and saw a black officer. It was Ian Daley, but Mike didn't know that; all he saw was the uniform. Mike began trying to explain who he was, but blood, not words, spit from his mouth. The officer seemed disgusted and jumped back a step. Mike heard the man yelling at him to put his hands behind his back.

Mike couldn't believe this. He felt sore and dizzy and like he might fall down. The officer was looking to cuff him! Then Mike had an idea: He flailed at his black parka, trying to open the jacket enough so the officer would see "something on my waist, my badge or something. I was just trying to identify myself."

The arm movements only alarmed Ian Daley, who, seeing Mike's handgun holstered on Mike's belt, drew his own weapon. Daley held the gun in his right hand, supported at the wrist by his left hand. His index finger rested on the trigger.

Mike pulled at the parka's zipper; he couldn't believe this. Then something was different. The officer must have seen Mike's badge and realized finally he was not one of the suspects. Mike heard the officer's voice: "Oh shit. Oh my God."

Mike took a step forward. But the officer just stood there. "He did nothing," Mike said later. Mike took another step, but walking was too much. Everything around him was spinning. "I don't remember falling but I remember being on the ground again." His head hurt, and he held the spot on his forehead that was bleeding the most.

He knew he was losing consciousness. "I just wanted to like sleep." He was alone again, struggling as he blacked out to fathom the unfathomable: How could this be?

While Kenny Conley was handcuffing Smut Brown, other officers arrived, including a patrol supervisor and a black officer who returned Kenny's radio, which had fallen during the foot chase. Kenny didn't know any of the officers—they were all from the immediate police districts while Kenny was far from his in the South End. He handed off the suspect to two officers who arrived in a police wagon.

Then he retraced his steps through the woods. He was checking for anything Smut might have discarded during the run, but he didn't find anything. He made his way back across the street and up the short hill to the fence surrounding the dead end. The scene surprised him. The area was all lit up. "The whole street was just lined up with cars."

He stood there and took it all in. Officers were all over the circle, including a bunch he knew: Dave Williams and Jimmy Burgio in uniform, and Gary Ryan and Joe Teahan, dressed in street clothes, from the gang unit. But what caught his attention was an ambulance, where paramedics were loading a black man strapped to a gurney into the back. The injured man was dressed in baggy jeans and a hooded sweatshirt.

"What happened?" Kenny asked.

"It's a cop," replied a security guard standing at the fence. The guard then recited the story already circulating around the dead end: "Hit his head on the ice."

PART II

True Blue

CHAPTER 9

"8-Boy"

When Kimberly Cox approached her husband in the acute care unit of the Boston City Hospital, her first words to Mike had a clinical purpose: to determine his level of responsiveness. She found Mike able to talk, but he was groggy and only "semi with it." Mike would try to speak, but he was unable to summon the words to complete a sentence. He complained about his head, with its swollen black mass, about feeling dizzy, about pain in his flank and in his abdomen. "He just looked very much out of it."

Mike also complained about his right hand, and Kimberly noticed his right thumb had ballooned. It was determined Mike had torn a ligament and hyperextended the thumb and finger—injuries that most likely occurred as he tried to break his fall.

Kimberly watched as the more than three-inch laceration on his forehead, still bleeding when she arrived, was treated and stitched. Nurses wiped off the blood caked around his swollen nose and mouth. More sutures were used to close the deep cut inside his upper lip, while the many smaller cuts and scratches were cleaned and bandaged.

Mike kept clutching his midsection, saying he felt as if he needed constantly to pee. Kimberly found a portable urinal and supported him. "I noticed that the urine was really dark." She sought out one of the attending physicians, showed him the urine, and asked that the doctor "dipstick it." The test showed traces of blood, hematuria. The doctors ordered further testing to explore the possibility of kidney damage and internal bleeding.

By 4 A.M., Mike was wheeled down the hall into the radiology unit for a series of examinations of his liver, spine, facial bones, and brain. The X-rays showed "no evidence of fracture" on his facial bones and nose, while the CAT scan showed that "the surfaces of the brain are clear." The results were favorable regarding Mike's neurological condition, although in the weeks and months to come, that would change.

In another bay, nurses were getting Jimmy Rattigan ready to be released. Rattigan had come into the hospital strapped to a backboard, wearing a neck collar; fortunately, neither he nor his partner was injured seriously in the crash with the gold Lexus. They were treated for bumps, bruises, and strained back and neck muscles.

Rattigan watched doctors and nurses attend to Mike; he saw Kimberly and Mike's mother arrive, and he saw other officers come and go. He could tell everyone was worried. "Michael Cox is one of the nicest guys I've ever met in my life, and one of the nicest guys I've ever worked with—always a gentleman, always says hello, never in a bad mood when you saw him. I was kinda worried for him, too."

Rattigan did pick up one tidbit about Mike's condition. "I heard before I was leaving that he might have been urinating blood—now that's definitely not a good sign."

Rattigan was right—blood in the urine was not a good thing. But unknown to Rattigan, Mike was going to have a lot more to worry about than traces of blood in his urine. Mike's concerns would soon enough extend beyond the physical to the metaphysical—a Boston police officer's expectation for justice was about to collide with the police culture of silence.

Having blacked out, Mike had missed the chaos that continued swirling in the compact cul-de-sac at the end of Woodruff Way.

By some estimates, more than twenty police cruisers from various departments, several ambulances, and dozens of officers ended up crowded into the dead end. The cruisers' lights sliced up the sky. "It looked like Christmas," one of the officers said later.

Joe Teahan and Gary Ryan were among the first to attend to him. Teahan discovered Mike alone on the ground behind Williams's cruiser, writhing in pain. "He was lying on a good-sized patch of ice," Teahan said. "He was hurt; he was bleeding." Teahan also saw the blood, with handprints, spread across the car's trunk. Kneeling down, he heard Mike moan, "I can't believe this shit. I don't need this fucking stuff."

Despite the cold, Teahan stripped off his sweatshirt and folded up his T-shirt. Gary Ryan used the undershirt like a bandage and tucked it under Mike's head. Teahan noticed that Mike had begun to shake. "He looked like he was getting cold."

Other officers, including a couple of munies, gathered around. They began calling for an ambulance. The requests broke an eerie stretch of silence on the police channel 3 after Mike's final scratchy transmission that the suspects were getting ready to bail. It was a vacuum that left the police dispatcher grasping for straws. "Where are we now?" he'd yelled. "Are we on foot? Could someone tell me? Are we on foot?"

The silence had finally ended with the calls for an ambulance. Even then, the requests lacked the necessary particulars—nothing about for whom, for what, or even where. "I need your location," the dispatcher said, stating the obvious.

The Woodruff Way cul-de-sac was described.

"What kind of injuries do we have?" the dispatcher asked.

The question hung in the air.

Gary Ryan then ended the suspense: "Officer with a head wound."

Many of the responding officers did not drive into the cul-de-sac itself. Two other members of Mike's gang unit, black officers Donald Caisey and Sergeant Ike Thomas, had circled in and had

driven down Mary Moore Beatty Way, the street below the dead
end and down a short hill from the fence. Mary Moore Beatty Way
was the street Smut Brown had run down after scaling the fence.
While they parked, they heard the radio call about an injured officer, but didn't know what to make of it. Caisey climbed the hill
ahead of Thomas and saw Teahan and Ryan huddled over Mike.
Hurrying over, he was shocked. "His injuries were very similar if
not exactly like injuries obtained when someone is shot in the head.
Clumps of blood coming out of his nose, out of his mouth, blood everywhere. Real thick blood."

Caisey leaned down. "Have you been shot?"

Mike heard the question, but no longer knew up from down. "I
don't know."

Thomas, meanwhile, was making his way through the hole in
the fence to gain access to the dead end. The shining cruiser lights
made it hard to see. But once he adjusted his eyes he noticed the
empty Lexus and three black men in handcuffs on the ground by
the curb. Then he recognized Ian Daley walking quickly in his direction. Thomas and Daley knew each other pretty well; earlier in
their careers they'd worked together in Dorchester at the C–11 station. They'd socialized on occasion and played basketball. But it
had been a few years since they'd hung out.

Daley motioned at Thomas. "Ike, Ike," he said.

"What's up, Ian?"

"Ike, you guys really ought to wear jackets."

Thomas was thrown. Jackets? It was like a non sequitur. He'd
just arrived and was quickly trying to take stock of all the activity
and figure out who was hurt—and Daley was in his face animatedly
insisting the gang unit's sartorial choices were lacking.

"I'm like, 'Okay, What are you talking about?'"

Daley could have given Thomas some context—told Thomas
that for a few scary moments he'd had his handgun trained on Mike
Cox, admitted he'd tried to arrest Mike. But Daley didn't do that.
"You guys really ought to wear jackets because some people don't
know who you are."

Thomas didn't have time for guessing games. He spotted Teahan and Ryan kneeling behind a cruiser and went over. That's when he saw Mike. From the vague radio transmissions, the sergeant had not gotten the sense an officer was badly injured. But Mike looked seriously hurt. Thomas asked Mike what happened. "He tried to talk," Thomas said later, "but he couldn't. Nothing was coming out." Richie Walker then came over and stood behind Thomas. Walker had himself just returned to the dead end. From his car, he'd run through the hole in the fence, banged up his knee after slipping, and then hustled after the suspect later identified as Smut Brown. He was the officer who retrieved Conley's flashlight. Walker asked how Mike was doing, but Thomas waved him off. The supervisor stood up and was asking out loud: What happened? What happened?

"We found him like this," said Ryan and Teahan.

Their response didn't answer the question.

Seconds later, Craig Jones was also hovering over his battered partner, a sight that took his breath away. From the point where he'd knocked down Tiny Evans on the left side of the cul-de-sac, Craig had run to the front of the cars and followed Richie Walker through the hole in the fence. "I assumed he was chasing somebody." Craig tripped going through the hole and slipped on the hill. Instead of joining the foot chase, he'd turned around and gone back to the dead end. He saw that Tiny, Marquis, and Boogie-Down were on the ground in handcuffs in front of the Lexus. To Craig, this was great news. Craig saw Dave Williams at the front of the Lexus. Excited by the successful climax to the long chase, Craig raised his hand and slapped Williams's—a congratulatory high-five between two towering black cops.

Craig had then noticed Tiny was yelling for him, squirming and trying to get to his knees. Craig went over and pushed Tiny down. He ordered Tiny to stay put. When Tiny didn't and said he needed to talk to him, Craig leaned down and punched him hard in the face. "He fell on the ground," Craig said. Tiny stayed put this time.

Craig was charged up. Cops and cruisers were everywhere, the sirens and lights a kind of sound and light show providing an exclamation point to the capture of shooting suspects. "My adrenaline was going." But the satisfied feeling was short-lived. Gary Ryan came over and told him, "Your partner is hurt." Craig followed Ryan to the rear of the cruiser. "Mike was a mess." The blood was all over Mike and splashed across the cruiser's trunk. The cuts, the bruises. Mike's head misshapen by the huge bump. "Never seen anything like it." Craig's soaring feelings had ended in a crash landing.

He knelt next to his partner. Mike seemed to respond to the familiar voice. In and out of consciousness, Mike finally made some sense on two fronts. One was professional: He began mumbling about the guns thrown from the Lexus, and seconds later the police radio crackled with his information. "The officer who is injured told me the suspects threw some weapons out on Itasca Street," reported one of the other officers standing there. The dispatcher asked for more details. "Uh, he's a little bit, uh, hurtin' right now," came the reply, followed by "He said two different locations."

Mike's second moment of clarity was personal: He asked Craig to call Kimberly. "He kind of moaned it," Craig said. Craig promised he would.

Craig and the others attending to Mike were growing impatient.

"Get an ambulance down here!" one of them yelled over the radio.

Seconds later: "What's the deal?"

"C'mon, hurry up!"

Hampered by the bottleneck of police vehicles, the paramedics arrived at 3:03 A.M. Sergeant Thomas watched as they worked furiously to stabilize Mike. "They were cutting his clothes off, taking his clothes off like it was a very, very serious injury," he said. "They were asking questions, like, 'Has he been shot?'"

Elsewhere, officers milled around, simultaneously checking on the three shooting suspects on the ground and curious about the damage done to Mike Cox. For their part, Dave Williams and Jimmy Burgio pretty much hung back ten to fifteen yards away in front of the Lexus. Williams, at one point, walked over to check on Mike, but Burgio stayed put. One thing on his mind was making sure he was going to get credit for an arrest. "You get written up for that, it looks great in your folder," he said later.

Mike was moved onto a stretcher. Thomas, the ranking officer, began making some decisions. He talked to Craig Jones about returning to Itasca Street to look for the handguns. He told Gary Ryan to ride in the ambulance with Mike, and he told Joe Teahan to follow in their car. He told them both to contact Mike's wife. He also made sure they took possession of Mike's equipment—standard operating procedure. Thomas got Mike's gun and handcuffs, but they couldn't find the radio Mike wore clipped to his belt.

Ryan and Teahan began looking around, a search that took Teahan down along the right side of Dave Williams's cruiser and past the passenger side where Jimmy Burgio rode. Teahan spotted Mike's radio on the ground in front of the cruiser by the fence. "It was kind of like in front of the car, but up like two o'clock," Teahan said.

No one was really paying attention to the significance of it, but the radio's location suggested that, following the first blows, Mike either fell or was pulled from where he'd initially stood at the fence to a spot near the front of the cruiser.

Thomas was not yet focused on the reality that he had a crime scene on his hands—the assault and battery of Boston police officer Michael Cox. Instead, he asked again Teahan and Ryan and other cops standing around, What happened? Did anyone see what happened? He got shrugs. He got silence. "I got no response," Thomas said.

Thomas might as well have been asking the meaning of life. No one broke rank to offer any information. Instead, an alternative explanation arose, originating in the circle around Mike and then working its way out, where cop after cop grabbed on to it like a raft in troubled waters. The story started with Teahan and Ryan, two fellow gang unit members. They began telling everyone Mike had slipped on a patch of ice. Even while insisting he had not seen anything, Teahan nevertheless was saying, "Michael had fallen and hit his head." He and Ryan hypothesized Mike had run from his cruiser and "hit the patch of ice and went flying."

Despite all the police training on the scene, despite the first impression of the paramedics, Donald Caisey, and others that Mike had been shot, the ice-slip theory quickly became gold. For those who'd beaten Mike or witnessed the attack, it provided cover for their crimes. For those who were not culpable but sensed trouble, it was a safe haven from having to consider wrongdoing by fellow cops. Mike's injuries were accidental—what could be simpler than that? The explanation was neutral and nonincriminating: textbook see no evil, hear no evil, and speak no evil. Even Craig initially went with the seductive but absurd concoction.

"I didn't want to believe what really happened, happened."

In denial or worse, Mike's coworkers launched the bogus story. That Sergeant Ike Thomas did not put the brakes on the rank speculation but instead allowed the smokescreen to gain traction was the first of a series of supervisory failures complicating right from the start any search for accountability.

"That's the only thing that I had to go on at that point in time," Thomas said later in self-defense, seeming to abandon altogether the most basic trait of any investigator: skepticism. Thomas even assigned Donald Caisey the job of writing the report about Mike's injury, and Caisey, notwithstanding his own initial skepticism, went with the storyline of convenience: ice.

The reality was that, in addition to the Lexus and the capture of the shooting suspects, Sergeant Thomas and the others now had a second crime scene requiring clear-cut steps to secure and preserve evidence: taping off the area around the fallen Mike, the cruiser, and the fence; photographing all those areas; seizing flashlights, boots, and clothing of those officers first to arrive to test for trace evidence in the crime lab; notifying immediately the command staff and Internal Affairs; taking statements from the officers at Woodruff Way.

But rather than consider that Mike had been mistaken for a fleeing shooting suspect and beaten to a pulp, Thomas and others in charge steered clear, pursuing instead a kind of supervisory avoidance of the obvious.

"It seemed believable to me," Sergeant David C. Murphy said later about his embrace of the ice-slip theory. Murphy was the second sergeant to show up at Woodruff Way, while Mike was still on the ground and everyone was waiting for the ambulance's arrival. He was the patrol supervisor from the Mattapan station. The high-speed chase had cut across Mattapan and ended on his turf, involving a number of his officers, like Richie Walker. Murphy's job was to help Sergeant Thomas sort out the situation. He had parked his cruiser down on Mary Moore Beatty Way and walked up through the hole in the fence.

Murphy sized up the scene as having three parts that required supervision: the injured officer, the damaged police cruiser driven by Dave Williams, and the shooting suspects. He asked a few questions about Mike Cox and took at face value the speculation about a slip on ice. Not his problem—injuries to an officer were the worry of the officer's supervisor, Sergeant Thomas. Then there was the damaged cruiser—it was from the Dorchester district. Not his problem again—the vehicle was the worry of his counterpart from Dorchester, a sergeant named Daniel Dovidio. Murphy instructed Dave Williams to radio Dovidio and tell him to come to Woodruff Way.

Murphy then dealt with the three suspects and the Lexus. He oversaw arrangements for transporting Tiny, Marquis, and Boogie-Down to the Roxbury station house for booking, for photographing the Lexus, and for towing the car. Smut Brown had already been taken in a separate police wagon to the station. Murphy then listened to several officers, including Ian Daley, who wanted to talk up their roles in the arrests. "Everybody wanted a piece of this," Murphy said later. Murphy himself joined the unabashed maneuvering for glory. He later told police officials that down below on

Mary Moore Beatty Way he'd helped capture Smut Brown—an em-
bellishment completely at odds with the fact that Kenny Conley
apprehended Smut Brown.

With a singular focus on the shooting suspects, Murphy walked
around and began shooing cops away from the scene. "He started
telling everybody to get the hell out of there," said Bobby Dwan,
who had reunited with Kenny Conley after Kenny's return to the
dead end. Said Murphy, "There was a lot of people there who were
just kind of milling around." Murphy also began spreading the
canard about Cox—telling officers who asked about Mike's condition that he'd fallen and hit his head.

Inadvertently, or worse, Murphy was aiding and abetting the
developing smokescreen hiding the true nature of Mike's injuries.
He was clearing the dead end of officers who either were eyewitnesses to the beating or had picked up information about it. While
paramedics loaded Mike into the ambulance, officers, instead of
being ordered to document their actions, were told to disappear
into the night.

By 3:15 A.M., the ambulance carrying Mike Cox slowly worked its
way out of the dead end en route to Boston City Hospital about six
miles away. Many officers, following Murphy's command, were pulling out. Craig Jones left to retrace his steps to search for the handguns. Richie Walker got into his cruiser to head back to the car he'd
stopped near the Cortee's and then abandoned to join the chase.
Kenny Conley and Bobby Dwan returned to their cruiser; by 3:30
A.M. they were gone. They headed first to the Roxbury police station.
Kenny wanted to retrieve the handcuffs he'd used to complete the
task Mike Cox had started—the capture of Smut Brown. From there,
they continued driving through Roxbury back up to their sector in
the city's South End, where the next call they took was about yet
another "suspicious person" on Washington Street. They found a
hooker standing alone in the cold and ordered her to move on.

While some left, plenty of police were still amid the chaos
of Woodruff Way—Sergeants Ike Thomas and David Murphy,

Donald Caisey from the gang unit, Ian Daley, and a slew of officers from Boston and other police agencies. TV news crews began appearing. Dave Williams and Jimmy Burgio climbed inside Williams's cruiser to await the arrival of Sergeant Dovidio, their patrol sergeant. To stay warm, they blasted the heater.

Dovidio, the third sergeant, then showed up; in short order, he trumped the supervisory mess already in play at Woodruff Way. His became the starkest display of disregard of duty. Dovidio was fifty-eight years old and nearing retirement. It was as if he wanted nothing to do with actual police work. Earlier, when the high-speed police chase went one way, Dovidio went the other. Even though several of his men, including Dave Williams and Jimmy Burgio, had responded to the shooting at Walaikum's, Dovidio drove back to the station. He said he had some paperwork to do. Besides, he said later, he was not obligated to get involved in the chase. "It didn't originate on my district."

But Dovidio did have to leave his desk once Dave Williams radioed about the damaged cruiser. The sergeant was not happy. He pulled up behind the Lexus, marched to where Williams and Burgio sat in the cruiser, and wanted to know where Burgio's cruiser was. When they explained that Jimmy's was back at the station, Dovidio demanded to know what the hell was going on. He began yelling about violating department procedures for teaming up without permission. Burgio tried to settle the sergeant down.

"What are you worried about? I just made a great arrest," he claimed.

Dovidio would have none of it. "The captain will have my ass." Thinking it over, Dovidio quickly devised a solution: He told Williams and Burgio they had *not* been in the one cruiser; instead, Burgio had been in his own cruiser. That was how they were going to write up their reports: There was not one, but two cruisers from the Dorchester station. He even pointed to a spot in the cul-de-sac where Burgio should say his cruiser came to a halt. No matter that the deception would create all kinds of confusion for investigators later trying to map out the scene. Dovidio had come up with an expedient way out for all of them—one that came with his imprimatur to lie.

The message was clear: Protect one another. Dovidio wasn't done either. Despite the flood of police officers, the sergeant, rather than naming names, was planning to say the only officers at the dead end when he arrived at 3:15 were the two in his charge. It was a fiction that helped clear the stage and create running room for those wishing to be invisible. In a third move, Dovidio decided he was going to try to see that Dave Williams and Jimmy Burgio were honored for exemplary police work during the wildest police chase anyone in the department could remember.

Dovidio, all by himself, was the embodiment of the culture of silence and cover-up that was kicking into gear on Woodruff Way. But he was not alone. "Bottom line," Kenny Conley said later, "is that no one took responsibility for that crime scene.

"The patrol supervisor [Murphy] from B–3 tried to say he caught the guy I cuffed . . . The anti-gang supervisor [Thomas] never really did what he was supposed to do. Lotta people lied that night. Believe me, Dovidio wasn't the only supervisor to neglect his duty."

After Mike's arrival at the hospital, while much of the city still slept, police were busy at two locations dealing with the chase's messy aftermath. The critical care unit at Boston City Hospital saw the comings and goings of some of Mike's coworkers, as they checked on Mike's condition and heard from doctors Mike had been hit with a "blunt object." No amount of wishful or deceitful thinking could turn an ice patch into the culprit.

Then there was the Roxbury district station, where the four shooting suspects were taken for processing and where a couple of dozen officers came and went as part of the post-chase debriefing. "There was a lot of activity in the station, a lot of people around," Craig Jones said. Given Lyle Jackson's numerous gunshot wounds, homicide detectives had joined the other officers who were congregating, either filing in from Walaikum's, the chase's starting point, or Woodruff Way, the end point.

Smut Brown and his three friends were searched, fingerprinted, and handcuffed to the wall in the booking area on the first floor. Smut was the only one to give police his true name. The other three offered aliases they'd used before: Tiny Evans said his name was Anthony Wilson; his brother Marquis said he was Robert White; and Boogie-Down became Darryl Greene. In evidence bags, officers logged their beepers, cell phones, necklaces and rings, Smut's $795 and Tiny's $707.

Smut hadn't been able to talk to the others and still thought Marquis was the victim of the police beating. Marquis was indeed hurting, but his aches came from the hit he took from the skidding cruiser. He was complaining he needed medical attention for his legs. Boogie-Down also wanted to see a doctor. "My right side of my face was scraped and bruised. My right hand was bruised. My lower back was bruised and my legs were hurting." The two were taken by police escort to Brigham and Women's Hospital.

In the second-floor "guardroom" a core group of cops was seated at desks to begin the paperwork. It was work with two tracks: one focused on the chase and doling out credit for the four arrests, the other focused on Mike Cox's injuries. Ian Daley had gotten the job of authoring the official incident report about the chase, known as a 1.1, a reference to the department's standard Form 1.1, while Donald Caisey of the gang unit continued working on the 1.1 about Mike's injury.

Police reports are supposed to be objective, reliable, and detailed, but interest in those principles—interest in what *truly* happened at the dead end—seemed lost as the guardroom was transformed into a creative-writing seminar. Officers huddled for brief chats about their competing versions of events, sometimes leaving the room for private talks, all the while keeping an eye on how the reports were stacking up.

"Everybody was trying to add their little bit to the report—'I'm so and so, don't forget me.' Things like that," Craig Jones said. The process was neither orderly nor pretty. Flare-ups erupted. Craig

and Ian Daley didn't hit it off, for example. Writing reports was
ordinarily an anathema, and Craig was put off that Daley was adamant about being in charge. He suspected that Daley saw being the
writer as a way to control the narrative to say he made one of the arrests. "I felt like he wanted the glory," Craig said.

But the others weren't about to let that happen. Craig main-
tained he'd arrested the driver. The towering Dave Williams then
claimed he and Jimmy Burgio arrested the two suspects in front of
the Lexus. Burgio even showed up at the station briefly to reinforce
the point. "It was my arrest," he said. "I wanted it." Richie Walker
then claimed he arrested the fourth suspect, Smut Brown. He acknowledged that another cop was there, but he didn't know who—a
"tall, white cop" was the best he could do. Craig chimed in the description fit Gary Ryan of the gang unit, and that was that. The
second cop became Gary Ryan. Mystery solved. No one bothered
to call Ryan to learn this was all wrong.

Instead, an overwhelmed Daley dutifully jotted down notes—
and, for the purposes of the report, not only was Kenny Conley suddenly Gary Ryan, but Richie Walker was the hero. "Officer Walker
never losing sight of the suspect," wrote Daley, "ran through bushes
and behind buildings and finally captured suspect." It was all part
of the misleading mess that Daley typed up in his three-page narrative. He did get some measure of revenge against Craig Jones; he
avoided giving Craig credit for arresting Tiny by not describing
Tiny's apprehension. Otherwise, it was all there—the key inaccura-
cies that Richie Walker captured Smut Brown, and that Williams
and Burgio had captured Marquis and Boogie-Down "after a brief
foot chase." The part about the foot chase was yet another fabrica-
tion; after they were knocked to the ground, Marquis and Boogie-
Down barely moved, except to wiggle out from under the cruiser.

But not every cop orbited the guardroom making sure Daley got
his name spelled correctly. Some went back to work—like Kenny
Conley and Bobby Dwan. Capturing Smut Brown was a career
highlight for Kenny. He'd never nabbed a shooting suspect before.
He was certainly as competitive as anyone and always gave it his

all—fighting for a rebound under the hoop, for example. But his game was more about grunt than glory, whether on the court or on the job. He'd never gotten a medal. His personality was not about ego. "I'm not like that," he said. With Smut, Kenny saw himself as making an "assist." He and Bobby had helped out the guys from the Roxbury and Mattapan districts, and at the shift's end, that's what he and Bobby wrote down in their log at their station.

Jimmy Rattigan was another cop who eschewed the ego mud wrestling. He and his partner swung by the station after their release from the hospital. They went upstairs and walked into the guardroom and found everyone going at it. Said Rattigan: "They were saying, 'I cuffed him,' 'No, I cuffed him,' and they were like, 'Well, I'm gonna take this one, you're gonna take that one.'" He saw Ian Daley, Craig Jones, Dave Williams, and Jimmy Burgio. Rattigan was disgusted. "I thought to myself, there's Mike Cox sitting in the hospital and these guys are arguing over who's gonna take an arrest. Me and Mark, we were like, 'Can you believe this shit?' and then we just left."

It wasn't as if they weren't talking at all about Mike. By dawn most everybody in the guardroom was aware the ice story was bogus. "When people were coming up to the guardroom," said Ian Daley, "they were basically saying who was the police officer that got, you know, beat up?" Said Richie Walker: "That was the topic of the conversation."

The talk, however, mostly circled around the beating instead of focusing on what actually happened—who did what and who saw what. There were moments, however—fleeting moments—where key cops, still in the heat of the night's events, made comments that started down the road of truth. Ian Daley was one who began heading in this direction. He pulled Donald Caisey aside at one point and told him he knew cops had beaten Mike Cox. Caisey, taken aback, pressed Daley for more information. What are you saying? he asked. "Cops did this to Mike," Daley said. Caisey said, Okay, who? Daley did not answer. That was as far as he'd go.

It was as if a paralysis had spread like a virus once everyone realized cops had beaten a cop. The ground was unfamiliar to them—a coworker, a brother, had turned out to be the victim of police brutality, and some cop or cops at the station had either committed the assault or witnessed it. Had a suspect been beaten—well, that was not so otherworldly. Many cops had seen or been part of an altercation where the bad guys got roughed up and sometimes worse. This, however, was not the more typical us-versus-them dynamic that lent itself to sticking together to gloss over the use of excessive force. This was radioactive, a beating that was turbo-charged with a complicated set of competing loyalties—to the individual person, to race, to the code of silence, and to justice.

Daley was apparently not sure what to say or what to do. He was tongue-tied, but he also wasn't alone in "showing leg," or offering a hint of information of evidentiary value. Jimmy Burgio, down in the first-floor lobby, walked over to another one of the gang unit officers and said, "I think one of your guys got beat up by mistake." But, as with Daley, no one followed up in earnest, and Burgio said no more. It became another potential lead lost in the paralysis. Then the most tantalizing tidbit came from Dave Williams. Outside the guardroom, he caught up to Craig Jones walking down the hallway.

"I think my partner hit your partner by accident," he said.

Craig stopped. His investigatory gears kicked in. He pictured Mike's bloodied head on the pavement. Where's Burgio's flashlight? he asked. Williams said Burgio probably had it with him. Where's Burgio? Craig then asked. He's gone, Williams said.

Burgio had already left the station, and no one went after him. No one got the flashlight. Burgio was "8-boy," police radio code for "nowhere to be found."

The answers were right there in the guardroom. But no one came clean—and no one took charge and insisted upon it. No one called a time-out on all the bobbing and weaving to demand better. No one pounced on the leads—the incriminating and suggestive statements that in the light of the next day, and in the days that followed, began to be taken back, spun differently, or flat-out denied. The media were expecting police reports to continue its coverage about the hair-raising high-speed chase, portraying the arrests as a hugely successful police moment, and the few hints at the truth were choked off by a toxic blend of cop ego and cop cover-up. Mike's beating was a public relations disaster that would only steal a great headline about departmental heroics.

Sergeant Thomas allowed Donald Caisey's flawed injury report to go through—a minimalist composition of twelve handwritten lines in one ungrammatical paragraph: "Officer Cox lost is [*sic*] footing on a puddle of ice causing him [to] lose his balance and fall forward striking his head on a marked cruiser. Officer Cox then fell backward on the ground striking the back of his head on the ground. As a result of this Officer Cox sustained head injuries causing him to lose consciousness for a short period of time."

No mention was made of all the blood, of Mike's cuts on his mouth, the three-inch cut on his forehead and other facial cuts, the stitches, the egg-sized hematoma, the bruising to his midsection, the hand and its torn ligament, the kidney damage. No mention was made of the truth everyone at the station knew by daybreak—Mike was beaten.

The result was this: an initial official record of the event that was false—a record that kept the story simple and singularly about police success and that postponed the negative about Mike. But the phony reports—mirroring Sergeant Dovidio's handiwork at the scene—were tantamount to a license to lie. Once the lies began, where would they end?

Lying in pain in one of the cubicles in the critical care unit, Mike had tried to be helpful when the doctors asked him about his injuries. "The doctor asked me a lot of times what happened to me, and I couldn't really tell him." He had been glad to see his wife when she arrived. He didn't expect his mother to come, but was not surprised she did. In his fog, he saw officers coming and going, but much of it was a blur. Sergeant Thomas, Donald Caisey—he recognized them. His partner, Craig, was there, and before the night was over, Craig's hand was put in a cast. It turned out he'd broken a finger either when he punched Tiny Evans or when he slipped on the hill. Mike heard Richie Walker in the room say he'd seen him running toward the fence after someone, and Walker's comments helped Mike remember for the first time he'd been involved in a foot chase toward the fence.

Most memorably, he heard Dave Williams come in and say, "I think cops did this." He'd heard his mother gasp, and his first reaction was to worry about her. "She began making statements like, 'Oh, God,' and I just remember wanting to have a conversation calming her down because it was bad enough. She hated the way I looked and thought that I was going to die right there. I just wanted to calm her down."

It was vintage Mike, worrying about what others were thinking.

Kimberly, meanwhile, was left speechless. She didn't know what to think, but soon enough she began smoldering inside. She kept returning her gaze to the large bump on Mike's head. "It was a huge mass," she said. To her, it was proof Mike had been hit hard. The idea made her furious: Mike had been attacked by other officers who'd crossed the line from reasonable to excessive force. They'd hit him and then run.

"They shouldn't have left him," she said.

Mike picked up on his wife's anger. But, of course, he was mostly in the dark, and his family was focused on his care, not getting to the bottom of his beating. The doctors wanted to admit him for further treatment. Kimberly had a different idea. She wanted to take Mike home. "I could watch him closer." By her calculation, her one-on-one care was better than, say, a ratio of one hospital nurse for every five patients. She was graduating in five months from medical school, and she'd done a rotation in neurology at the New England Medical Center. She was confident she knew what to look for.

"I just felt I could give him better care at home."

So while daylight spread over the city, Kimberly brought Mike home to the two-family house on Supple Street in Dorchester owned by one of Mike's sisters. The light outside bothered Mike and made him squint. Voices made him cringe. His head pounded. Walking unsteadily to the house, he felt the world spinning. He could not think straight.

His sons, Mike Jr. and Nick, were waiting, wondering why their parents weren't home when they awoke. "They really looked up to my husband," Kimberly said. The boys sometimes talked about growing up to be a police officer like their father. "He's this big guy," she said. "Daddy, you know, is invincible." The boys had never really seen their father sick or off his feet. "All of a sudden, he's been knocked down."

Nick, who had turned five earlier in the month, hung back while his father was helped into bed. Then he slowly walked into his parents' bedroom—and he froze. He turned and ran quickly from the sleeping giant—from the man who was supposed to be his father but whose misshapen and monstrous-looking face was unrecognizable.

Nearly a week passed before Nick ventured back.

CHAPTER 10

No Official Complaint

Sometime during Mike's first night home he bolted up in bed. It was a sudden and nearly violent movement. Kimberly awakened immediately. She saw that Mike was soaked in his own sweat.

What's wrong? she asked.

Mike didn't respond. His shoulders shuddered, as if trying to hide something.

Then Kimberly heard: Mike was crying.

Mike? she said.

Mike wasn't sure what was going on. He felt frightened to the bone, a feeling that was crystal clear, even though he'd come home woozy from a combination of his injuries and the medications doctors had him taking—painkillers and antibiotics. Fear raced through his body like an electrical current, and he couldn't seem to get control of it.

Mike, Kimberly said. What's wrong?

When he first sat up it had all seemed so very real. But then he heard Kimberly's voice and began to realize where he was—in bed next to his wife. The house was quiet. He'd had a nightmare.

Kimberly asked about what. But Mike wouldn't say. More wakeful, he grew self-conscious about crying. Embarrassment replaced the fear. The two emotions were foreign to a man known for his poise and courage.

What was it? Kimberly asked again.

Mike still wouldn't say. "He just didn't want to talk about it," she said.

The nightmare wasn't a one-time occurrence. It came back the next night and again most nights in the weeks to come. In it, men in blue uniforms were after him. They were Boston policemen; Mike recognized the uniforms. But he didn't know who they were; they were faceless. They invaded his house and were usually armed. They opened fire as they came toward him. Mike had his gun, but he was one against many.

Eventually Mike talked to Kimberly about the dreams. "The theme is usually the same," Kimberly said. "Our house is being stormed by several policemen with guns and he's shooting back. They're shooting and they're killing us."

Kimberly's commuting to Philadelphia for medical school had slowed down during her fourth year. She'd arranged to do her training in Boston area hospitals and was mostly home. The flexibility was fortuitous. Once Mike got hurt, she was able to stay with Mike to oversee his care and work on comforting the boys.

To protect Mike Jr. and Nick from the truth, she actually adopted the official explanation for his injuries. The next day she told them their father had banged his head after slipping on ice. Five-year-old Nick, in particular, was scared his father was going to die. She reassured him he was not. But Nick stayed frightened and easily upset. Kimberly found herself putting Nick to bed early and sitting with him until he fell asleep.

The boy's fright tore Mike up. "He wouldn't talk to me." Mike was bedridden and helpless to do much of anything his first week home. It was like he was trapped in a thick fog. He hated to move. The slightest turn caused the room to rock. "I couldn't get up quickly, or turn my head quickly," he said. "I would get dizzy and fall." The best chance to keep the world still was to move in slow

motion. To go to the bathroom the morning he got home, he shuffled across the bedroom, and that's when he saw for himself what
he'd overheard at the hospital: His urine was "very dark, dark, with
strands of like red in it."

His mind was off speed too. He couldn't find the words to complete a sentence. He would start, and then the words seemed to slip
through his fingers. When he wanted to call out to one of his sons,
he couldn't. It was like his mind was stuttering.

"Just to remember my kids' names was, like, a struggle."

It was scary, and Mike was a wreck. Neither his body nor his
mind felt like it belonged to him. Following an examination, a
neurologist concluded that Mike had post-concussive syndrome
and post-traumatic vestibular vertigo, medical-speak for what was
causing his splitting headaches, dizziness, memory loss, and cogni-
tive difficulties.

No matter how much he slept, he continued to feel sluggish. "I
went out to a doctor's appointment, came back, and lay back down."
In a way, not much had changed since right before he blacked out
near the fence: He felt tired and just wanted to go to sleep. The last
thing he wanted to do was talk about what happened.

Mike may not have felt much like talking, but others did. His
mother, his sisters, and Kimberly were boiling mad about the beating and right away wanted Mike to do *something*, even if they didn't
know specifically what that meant. Within days of his coming
home, his sister Lillian wanted to photograph Mike's face to document visually the extensive swelling, bruising, cuts, and bumps. She
and the others told Mike that the police department was going to
try to sweep the beating under the rug unless he took action.

Mike began hearing the same message from beyond his immediate family. Leaders from the local chapter of the NAACP and
the Nation of Islam called the house. Mike had never been active
politically or religiously. "I was amazed how many people got my
phone number," he said. The groups had heard about Mike's beating
through talk on the street, and they wanted him to do something

about it. "What do you want to do?" they'd ask. One day Mike took a call from another black cop on the force. Mike didn't know the other cop very well, but that didn't stop the caller from getting into Mike's business. "He's asking me how I'm doing," Mike said, "and then his tone changed and he said, 'I've known you for a while and I've always respected you, but if you don't do something I've lost all respect for you, as a person, as a black man, as a police officer.'"

Mike didn't want any of it. To him, it was all noise and static. The ground beneath him was already unsteady—literally—and he was having enough trouble finding his footing. "I was just happy to be alive," he said. "I'm just trying to deal with the day-to-day, with my injuries." So he refused to let his sister photograph him. He rebuffed any other calls to action—there'd be no protests, no press conferences.

Instead, he told his wife, mother, and sisters it would all work out. They fired back that Mike was being naive. "They were like, 'Why are you so trusting? What's wrong, can't you see?'" But Mike would not budge. They didn't understand cops. They didn't understand the split-second decisions of a high-speed chase. They couldn't put themselves in the beaters' shoes as could Mike. "Maybe, you know, they thought I was the murderer," he said. "So maybe trying to arrest me was justified."

Mike's first instincts were true blue. The severity of the thrashing notwithstanding, Mike got that it had been a terrible mistake. Unlike his family, he didn't see making a federal case out of it. Friends from the gang unit came by the house the first week to check on him. He wasn't up for talking much, but he listened, and Craig Jones told him what Dave Williams said about his partner, Burgio, messing up. Mike heard from Dave Williams, and Mike thought Dave sounded "very apologetic." From others he heard gossip the brass was giving those responsible some time—a grace period, of sorts—to come forward before any kind of intense internal probe was begun. The tidbits gave Mike the idea this was going to get resolved and settled in a way he preferred both personally and as part of the fraternity himself—quietly and within the organization. "I felt this loyalty to police in general." He was optimistic, knowing full well police officers tended to protect another suspected of misconduct. But he also believed this went beyond any unspoken code of silence. When the victim was one of your own, it was a different ball game.

Mike was figuring that within days he'd hear from the cops who'd beaten him. He was counting on an apology. "I expected the individuals to come forward and say what they had done." They'd get disciplined in some fashion. Then they'd all move on.

Mike didn't expect his wife, mother, and sisters to understand any of this.

"In the beginning I had a lot of faith," Mike said.

In law enforcement it's a well-known truism that the chances of solving a crime diminish the longer a case goes unsolved. For one thing, witnesses have time to think about what to say or to decide not to say anything at all. Offenders have time to work out the wrinkles in their cover stories. "The best moment for justice is right away," one prosecutor said.

In the Cox case, some Boston police officials may have been hoping the department's low-key response to the beating would result in a quick and quiet resolution that kept the matter largely in-house. They may have figured that given the unique circumstances—cops beating a cop—it was reasonable to expect the offenders to come forward. Mike had thought as much. They were wrong.

In the first days after the beating, instead of launching a full-blown Internal Affairs inquiry, commanders in the field put out the word that officers in Roxbury and Mattapan who'd participated in the chase and were at the dead end had to file a so-called Form 26 report. The officers were to document what they'd done and what they'd seen.

For two weeks the Form 26s trickled in—and the false notes struck the morning of January 25 in initial reports played on. Nearly sixty officers prepared reports—and not a single officer saw or knew anything about Mike's misfortune. The ultimate see-no-evil filings were done by the core group of officers who, along with Mike and Craig, arrived first to the dead end: gang unit officers Gary Ryan and Joe Teahan, Richie Walker, Ian Daley, Dave Williams, and Jimmy Burgio.

Ryan's report said he and his partner, Teahan, rode in "the fourth m/v [motor vehicle] on the scene," where they found "Michael Cox lying on the ground." That was about all he wrote; Ryan said he didn't see anybody else and had no idea how Mike got hurt.

Richie Walker, according to his written report, had not even seen Mike Cox: "I learned via my portable Boston Police radio that there was an injured officer at the location where the pursuit had ended."

Ian Daley drafted a brief, handwritten account in which he at least acknowledged a Cox sighting: "At the conclusion of the pursuit, Officer Daley did observe P.O. Cox laying on the ground bleeding." But Daley provided no meaningful details.

Dave Williams used a typewriter to fashion his report—all sixty-four words of it. He said he and Burgio were "involved in a foot chase," suggesting he was in no position whatsoever to see Mike. Williams even got Mike's name wrong, typing, "I was unaware of any injury to P.O. Richard Cox until later."

Jimmy Burgio, in the few sentences he prepared, avoided any mention of Mike by focusing singularly on his moment of putative glory: "Myself and P.O. Williams engaged in a brief foot pursuit ending with the arrest of two suspects."

Their supervisor, Sergeant Dan Dovidio, then extended the cloak of cover. He filed a report saying no other officers were at the dead end when he arrived, except for Williams and Burgio. Separately, he then sought honors for the two officers. He typed a "Recommendation for Commendation," writing, "Officers Williams and Burgio are worthy of recognition and should be commended for their excellent performance that without doubt instills public confidence to victims of violent crimes."

The collective exercise in evasion did little to shed light on the beating. The leaders of Mike's gang unit were growing restless. The unit's commander on January 30—five days after the beating—fired off a single-spaced, three-page memorandum forcefully calling on the department's top brass to start an Internal Affairs investigation. "The most disturbing aspect of this," the commander wrote, "is that not only was Officer Cox assaulted, but he was *left* on the sidewalk." The commander stressed, "as of the writing of this report, no officer has come forward, in spite of the fact that there were numerous officers involved in this initial vehicle and suspect pursuit."

Four days later, the *Boston Herald* ran the first story mentioning Mike in connection with the high-speed chase and capture of four murder suspects. The tabloid's early "bulldog" edition hit the streets not long after midnight and was usually read by the cops, firefighters, cabdrivers, and anyone else working in the dark. The story said Mike had been injured and the department was investigating whether he'd been beaten.

Later in the night, the telephone rang and awoke Mike and Kimberly. Mike turned carefully and grabbed for the receiver by their bed. He heard a voice grumbling. The voice was unfamiliar, and Mike had to hold the receiver away from his ear when the grumbling grew into a primal scream. The scream spread through the room. Then it stopped. The caller hung up. Mike put down the receiver. He was puzzled, but didn't think much of the weirdness. He and Kimberly tried to get comfortable again.

Then the telephone rang again—and again. Each time Mike picked up the receiver to hear the same animallike scream. Then the caller hung up. Toward dawn, Mike picked up the receiver, bracing for the scream, but it didn't come. Mike said, Hello? The caller asked for someone by name, but uttered a name that made no sense. "It was a nonsense name," Mike said. "It was a name I couldn't even pronounce."

Mike asked the caller, Who?

"You asshole," the caller yelled. "Fuck you!"

The line went dead, and that was it. The calls ended. They didn't belong to one of Mike's nightmares. The telephone calls had been real, and they left Mike and Kimberly bleary-eyed. But they weren't going to puzzle too much over a wrong number or sick prank, not with all they had going on in the family, given Mike's condition.

It was Friday, February 3, nine days since Mike's beating.

Later in the morning a friend of Mike's called to tell him about a story he should read in that morning's *Herald*. The story was on page 16, and it carried the headline "Alleged Beating of Undercover Cop Probed." It was a brief account—289 words long—reporting that the department was looking into the possibility that "an undercover police officer was beaten by other officers at the height of a chase following a shooting last week."

Mike read the story carefully. He noticed a mistake right away; he worked in plainclothes and wasn't an "undercover" cop. The mistake didn't matter at this point. What mattered was this was the first public disclosure that Mike had been a casualty in what so far had been heralded in the media coverage as a night of sterling police work.

Mike read on: "Officer Michael Cox, 29, a member of the Anti–Gang Violence Unit, suffered kidney damage and head wounds in the Jan. 25 incident, which occurred as police pursued four suspects for a shooting at a Roxbury eatery, sources said."

The department's spokesman was quoted as saying, "This is serious."

Mike found himself thinking about the crank calls. He could hear the caller's voice in his head and it made him feel queasy. The calls were clearly connected to the story. He might be reading the story for the first time at midmorning, but Mike knew that cops working the overnight shift often grabbed the two morning papers, the *Herald* and the *Boston Globe*.

The caller, Mike decided, was not random, a nobody—he was a cop who'd read in the *Herald* that the department had started looking into the beating and that Mike was talking. He would never be able to prove it, but he knew it in his bones. Mike felt a panic. The newspaper story followed by the middle-of-the-night "Fuck you."

Juxtapose the two, and Mike knew the call was a warning: Keep your mouth shut.

The story itself presented another puzzle for Mike. In it, police sources were quoted saying they were trying to sort out what happened. One was quoted saying Mike "remembers being in pursuit, he remembers being struck, and that's all he remembers. Obviously something happened. But if he doesn't tell us, how are we going to know?"

If he doesn't tell us, how are we going to know? Mike reread the quotation. It was absurd, he thought, flat-out absurd—the notion only he had the key to the truth.

Then came this: "We have no official complaint yet. Michael has not come in and said he was beaten up."

No official complaint? thought Mike. The notion that police investigated violent assaults only after the victim filed a formal complaint was flat-out absurd. "Hogwash," Mike said. The department was making it sound like the ball was in his court—to both pursue the case and solve it.

None of this sounded good to Mike. The story in the city's other daily newspaper, the *Boston Globe*, only added to his anxiety. Like the *Herald*, the story reported police officials were "trying to determine how plainclothes officer Michael Cox was injured in the line of duty last week." But comments by the spokesman were, once again, misleading if not outright false. "We're not sure—he's not sure—how he was injured."

Mike was confounded. The two newspaper stories were like a punch in the stomach. The way he read them, the message was at best mixed. Officials were saying his injuries were serious but they didn't know how he got hurt. They were flummoxed.

Then came this: "There is no assumption of any wrongdoing yet."

No assumption of wrongdoing? Mike couldn't get past that line. It was nine days since he'd been beaten, and everybody knew he'd been beaten—mistakenly, perhaps, but he was beaten, and the beating was overkill, a case study of excessive force.

"Everybody and their mother knew about it," Mike said. Yet there it was, officials telling the public, "There is no assumption of any wrongdoing." This did not sound like a department determined to get to the bottom of the beating of one cop by other cops.

For the first time, Mike wondered what was going on. One thing, he had not heard anything directly from the police commissioner. Paul Evans had not visited the house or called to ask how Mike was doing. Evans had not issued any clear signal inside the department that the brutal beating broke all the rules—written or unwritten—for which he was demanding accountability. In the newspaper stories, Evans was not even quoted; he'd let his spokesman handle what was at once a deadly serious matter and a potentially huge embarrassment for the police. The commissioner was certainly busy with other ongoing embarrassments—most notably the botched drug raid that had left an elderly Dorchester minister dead. The day after Mike's beating, Evans had had to stand before the reporters to announce the suspension of one lieutenant and reprimands of two supervisors. Then, after announcing the disciplinary measures, Evans faced criticism he'd done "too little, too late." Meanwhile, lawyers for the city and the minister's widow were locked in sometimes nasty negotiations to settle her wrongful death claim.

The commissioner apparently did not have time for Mike Cox. And his remoteness, along with his spokesman's wishy-washy comments, stood in sharp relief to all the public concern about police brutality fifty miles down the road in Providence, Rhode Island. The videotape showing an officer kicking a black man in the stomach during a melee after a concert was a big, ongoing story. The police chief had gone public with his concern and condemnation of the apparent misconduct, and he was soon joined by the city's mayor, Vincent "Buddy" Cianci. "Let the chips fall where they may," Cianci told reporters. "We will not tolerate excessive force. We will not tolerate any brutality."

When the stories about Mike ran in the *Herald* and *Globe*, Mike's family jumped all over them. They saw the stories as clear-cut evidence supporting the point they'd been making—the police department was in cover-up mode. They said, We told you so, Mike, you have to do something! Then lawyers began calling the house, despite the unlisted number, to discuss with Mike the possibility of legal action. "I was amazed about how many people had my phone number," he said. Mike was appalled by the unsolicited calls. No, he said, despite his family's protestations. No. Even if he'd begun to wonder.

His family persisted. To make them happy, Mike agreed to meet with an attorney one of his sisters had come across on her own. His name was Stephen Roach, a forty-five-year-old civil trial attorney. Roach had just struck out on his own, teaming up with another lawyer to start his own firm downtown. He had been practicing law for just over a decade, competently but without fanfare. He was not well-known in the halls of power or in the media as one of the city's go-to lawyers who could make things happen in the corridors of justice. In Boston, it was always said that personal connections and who you knew mattered—in business, in politics, and in law. Roach was not a member of this elite club of insiders.

Roach was originally from the town of Houlton in northeast Maine along the Canadian border. He came to Boston to attend Boston College, graduating in 1973, and began studying law in Boston at Suffolk University Law School in the fall of 1979. Roach apparently liked a full plate; he held a full-time job while attending law school. He was intense and indefatigable and, once he got a taste of law school, displayed a streak of feisty litigiousness. The year he began law school, Roach and his roommate got into a beef with their landlord. The bathroom ceiling leaked. Pieces of rotted wood and plaster came loose and fell on them. The landlord ignored their demands to fix it, so Roach sued him. He brought a small claims action in the city's housing court, seeking reimbursement for $750 in rent. He and his roommate won. The landlord appealed, and the two sides negotiated a settlement for $400. It was the kind of landlord dispute most tenants only grouse about. Not Stephen Roach. For him, bring it on.

Roach showed up at the Coxes' on Supple Road in Dorchester

in early February. The meeting did not last long. "I just wanted him to leave as fast as possible," Mike said. "It was like, 'Hi, nice to meet you. I have a headache. Can you please leave?'" Mike had met with Roach mainly to placate his family, but the meeting did last long enough for Mike to realize Roach was an outsider. "He knew nothing about the Boston police," Mike said. "He had not worked for them, and he didn't seem to be part of that culture."

For Mike, this was good. Roach didn't owe anyone anything. "He seemed like a safe outlet." On his own, Mike had privately begun to question the low-key nature of the department's response. Maybe his family was on the right track. Maybe the "grace period" was not so much time to allow the wrongdoers to step up as to enable a cover-up to take root. Maybe the delay was to see whether Mike was going to push this thing; if Mike didn't, then maybe the debacle at the dead end just goes away.

Then there came a second call several nights later, followed by a third. "Virtually every night," Mike said. Sometimes the caller didn't say a word, other times the caller screamed, and other times he yelled, "Fuck you." Sometimes Mike lifted the receiver and left it on the floor. "An hour later I might put the phone back, and ten minutes after that the phone would ring." Mike became convinced the caller was a cop using blunt force—the linguistic equivalent to a nightstick or flashlight—to keep him down and silent. But if that was the intent, the harassing calls worked to an opposite effect on Mike: as a wake-up call from his deep slumber.

"It helped me to focus," Mike said. "This was *not* just gonna go away."

By the second week of February, investigators for Internal Affairs were working in earnest—an effort that began only after the initial newspaper accounts about Mike ran on February 3. The news stories might have been circumspect, but they had signaled the word was out. The department could no longer put off pursuing a formal look at the incident—and the first order of business was a sit-down with Mike Cox.

Mike arrived at police headquarters in downtown Boston late in the afternoon of February 9 and rode the elevator to the fourth-floor offices of Internal Affairs. For the division, the Cox case—officially known as Case #2795—was hardly standard stuff. When it came in, for example, investigators were checking out an officer who'd failed to report for duty on New Year's Day and apparently never called in sick. In another new case, investigators were sorting out who did what in a car accident involving a police officer and a retired city resident. The retiree complained the officer, while writing out a ticket, was abusive, yelling, "Fuck" and "Fuck you" at her.

In contrast, Mike's case was the kind of complicated and radio-active mess few investigators would want to touch—and, in the end, it was handed off to a relatively inexperienced investigator. Sergeant Detective Luis Cruz had only worked in the division for about a year. But it wasn't only his short service that stood out. Cruz had his mind elsewhere. The ambitious officer was wrapping up law school and looking to graduate in June. He also had been trying to get out of Internal Affairs. He wanted to work at the police academy training recruits. With a transfer imminent, Cruz nonetheless took the helm.

Mike's interview marked his first formal talk with investigators. His appearance actually may have surprised those in the department hoping Mike's fifteen days of silence signified he was going to do nothing. Indeed, Mike had been fielding regularly what he interpreted as messages to go this route. Some were crude—the crank telephone calls at night, for example. Others seemed more subtle. More than once, Mike listened while someone he knew on the force shared a story about being mistakenly roughed up by other cops. One was a fellow officer in the gang unit named Fred Waggett. "He had been hit with a baton before," Mike said. "The guy apologized and he let it go." Mike liked Waggett and was not offended by what he took as the theme to Waggett's first-person tale: Silence is golden. "He was giving advice he thought was legitimate," Mike said, "and he was sincere in how he thought I should handle the situation."

The truth was no one really knew what to expect from Mike
Cox in the aftermath of the beating. He was a quiet man who
guarded his privacy. Few on the force knew him besides his partner, Craig. Mike might have liked it that way, but the privacy came
with a price. Indeed, going back to when he was a boy, his reticence
was often misinterpreted. When Mike sat mute while his adviser
at Milton Academy accused him of smoking pot, the adviser con-
fidently took Mike's failure to speak up as confirmation. Now fifteen years later, Mike's silent ways were still being misconstrued.
Colleagues who came to see him thought his low profile meant he
was going to sit tight.

The week before the sit-down, for example, the department had
issued a press release announcing Mike's promotion to sergeant—
along with sixty other new sergeants, twenty lieutenants, and eight
captains. Even though the promotion was long in the works, Mike
immediately heard talk that the promotion was his payoff for not
being pushy. "The rumor was they're gonna take care of me, not to
worry." Mike should not have been surprised by rumors based on
false interpretations of his silence, but he was.

"I'm like, 'What are you talking about?'"

In fact, Mike's showing up at headquarters to see Sergeant
Detective Cruz did not constitute a turnabout of any kind. He'd
always wanted accountability—and nothing less. If he'd said little
to anyone about his expectations, it was due to his nature and his
injuries; headaches, for one, plagued him. The start of the Internal Affairs probe—however belated—was a good sign to him. Two
weeks had gone by, and the offenders had had their chance to come
clean. It was time for Internal Affairs to turn up the heat. "I had
family members telling me nothing would happen, but I was sure
they would get to the bottom of it," Mike said. "I believed that
wholeheartedly."

The interview with Cruz was taped. Mike's recollection about
what happened to him at the fence was still scant, but he tried his
best to be helpful. He was adamant about some points and wrong

about others. In the report he wrote for the interview, for example, he asserted clearly the ice-fall story was fiction. "I did not slip on ice or any other substance." But he incorrectly told Cruz that when he first ran from his cruiser toward the fence, he was chasing Smut Brown and a second suspect. The mistake was one Mike was never able to resolve; a faulty memory told him he'd run after two suspects.

Mike provided other new information—bits and pieces of the night's events that were slowly returning to him in the two weeks since he'd been home recuperating. He recalled that before he lost consciousness a white man wearing black boots kicked him in the face. He couldn't describe the man's features, however, or recall whether he was in uniform. "It's possible that I might later on remember." For now, that was it.

Mike offered one other new lead—another moment that had come back to him. He described standing, blood-soaked, behind a police cruiser, when "I see a black officer." The officer, he recalled, was yelling and trying to arrest him. He wore a uniform and had a slim build. "I know he's smaller than me," Mike told the investigator, but that was all he came up with.

The taped interview ended at 5:15 P.M. It had not lasted even thirty minutes. Mike still suffered from huge gaps in his memory. But even if he could not identify any of his assailants, he felt he had provided investigators with some good leads about one of the men who'd beaten him and about another officer who then tried to arrest him. Mike promised to pass along any other details—if and when they came back to him. The short interview left him feeling exhausted. Mike headed home to rest.

The next week, Mike ventured out to attend the ceremony honoring the eighty-nine newly promoted officers. During the event, Police Commissioner Paul Evans spoke to Mike for the first time since the beating. The private conversation amounted to a pep talk. Evans asked how Mike was feeling and encouraged him to get better so he could return to work soon. He told Mike not to worry about the "incident" and that "he would take care of it." They were encouraging words that Mike wanted to believe.

But it wasn't so clear whether the case was a high priority to Evans yet—or ever would be. Three weeks had now passed, and Evans had still not spoken out publicly about the beating. He'd certainly had the chance. The ceremony itself presented the latest opportunity. It coincided with Evans's first anniversary as "Boston's top cop," and the local newspapers used the occasion to write stories recapping his first year. In interviews with the *Boston Globe* and *Boston Herald*, the commissioner talked about the highs and lows. On the positive side, he noted the streets were safer as a result in a sharp drop in the city's crime rate, a decline he credited to putting more cops on patrol and forging alliances with neighborhood and religious leaders. The low point, he then said, was the death of the retired minister Accelyne Williams during a botched drug raid in Dorchester—a senseless tragedy that had been a headline story throughout the year.

Noticeable by its absence was any mention of Mike Cox. The Cox beating—one of the worst cases of police brutality in modern times—was a senseless assault that so far had only barely made the news, and the commissioner wasn't drawing attention to it.

It was important for Mike to attend the ceremony, but it wasn't easy. He didn't want to talk to anyone about the beating, and he felt people were staring at him. Some even seemed to be avoiding him. But Mike was proud of making sergeant, a rank he'd earned. The police world was still his world. And he knew others who'd won promotions. Mike was glad for them too. Diana Green, for one, also made sergeant. Mike had gotten to know "Dee" Green on the job working in Roxbury and Mattapan. She was originally from the South and had overcome a lot—childhood scoliosis as well as her father's accidental death—to become one of the top performers on the anti-crime unit. Like Mike's anti-gang unit, the assignment was elite, high-powered, and high-pressured. Dee Green was popular, a big-hearted cop who, following the beating, sought Mike out and suggested, gingerly, that talking with a therapist might be helpful.

"I don't really believe in that kind of stuff," Mike said. But Mike appreciated her interest and considered Green a trusted friend on the force. Going to the ceremony was a chance to see her and others whose police work he respected.

The February 14 ceremony actually turned out to be one of the few times Mike had left his house for something other than an appointment with either his primary care physicians or any number of specialists. Mike continued to see a battery of doctors for his headaches, fatigue, dizziness, and unsteady gait. He saw a neurologist for his "post-traumatic amnesia," and his doctors had another MRI done of his brain that, fortunately, "did not suggest intracranial bleeding or contusion." Concern about his "left flank discomfort" and "persistent hematuria," or blood in his urine, resulted in more blood tests and new examinations to assess possible damage. The day after his promotion to sergeant became official, Mike was on his back in another medical office while a urologist inserted a thin instrument called a cystoscope into his urethra and carefully pushed it up into his bladder. It was invasive, but the cystoscope allowed the doctor to look directly into his bladder into areas that typically did not show up well on X-rays. The procedure did not reveal any abnormalities in his bladder, urethra, or prostrate— which was good news. The bad news was Mike began peeing bright red blood and felt pain; the cystoscopy had caused a urinary tract infection. To his daily high dose of Tylenol and other medications, the doctor added antibiotics. It sometimes felt to Mike that he was still getting beat up.

One weekday near the end of February, Mike was walking down the hall at the police academy complex located in the city's Hyde Park neighborhood. He might have been off the street due to his injuries, but that didn't mean he was off the job entirely. Newly promoted officers had to attend classes. "I was still required while I was out injured to go to the police academy for superior officer training," he said. It wasn't heavy lifting, and Mike found time to go "in between doctor appointments."

It was late morning, and Mike was headed toward the cafeteria. "Hey, Mike, how are you doing?"

Mike looked up to see who was talking to him. He saw a black officer walking out of the cafeteria. "Mike," the officer repeated, "how you doing?"

Mike studied the officer, but drew a blank. The man stood a good six inches shorter than Mike. He had a slim build. The face looked vaguely familiar, and he was certainly acting friendly enough. But Mike could not place him.

"What's your name?" Mike asked.

"Ian," the officer replied.

"Ian what?"

"Ian Daley, sir."

It didn't help. The name meant nothing.

The exchange ended awkwardly. Mike headed home. He worried he was supposed to know the officer but was unable to because of his clouded memory. Then at home that night he thought more about the encounter and experienced an epiphany. He realized suddenly where he'd seen the officer named Ian Daley.

"It came to me about his face," he said. Daley, he realized, was the officer who'd tried to arrest him at Woodruff Way after he'd been beaten. It was as if Mike's heart skipped a beat—he'd solved a piece of the puzzle that muddled his mind. He needed to contact Internal Affairs. They'd surely want to follow up on his breakthrough.

CHAPTER 11

Can I Talk to My Lawyer?

When investigators for the Internal Affairs Division of the Boston Police Department sat across from police officer Ian Daley on the morning of March 2, much of the nation was riveted by the daily Court TV broadcasts of the trial of O. J. Simpson, the former football star charged with the murder of his ex-wife and her friend. It was the most highly publicized trial ever—all day with O.J.—with nearly 24/7 coverage. Meanwhile, in Boston, the police probe into the beating of Mike Cox was a local matter unfolding in quiet and secrecy, with barely a mention in the media.

The investigators set up in IA's interview room on the fourth floor of police headquarters, located on Berkeley Street a few blocks from the Public Garden in downtown Boston. The small room looked out onto a narrow side street and a Bertucci's pizza restaurant down the way. The lead investigator, Sergeant Detective Luis Cruz, was joined by Lieutenant Detective Jim Hussey. Hussey was in the process of taking over the inquiry from Cruz, who would soon be off the case with his transfer to the police academy.

Hussey was feeling terrible about Mike Cox. He'd met Mike at the academy in 1989 when he was an instructor and Mike was a new recruit. He'd followed Mike's start on the force and knew about Mike and Craig Jones's feats on the street. He'd learned recently from Mike that Daley was the officer who'd tried to arrest him after the beating. Mike told him a chance encounter with Daley at the academy had triggered his memory. Then, in interviews with other guys in Mike's unit, Hussey learned that after the beating, Daley told Sergeant Ike Thomas that officers working the streets in plainclothes needed to wear "jackets so people know who you are." To Hussey, it was certainly looking as if Daley knew something about what happened on Woodruff Way.

Daley arrived for the interview with an attorney provided by the police union. Although he'd been born in England, Daley, now twenty-nine, had grown up in Boston. Like so many officers involved in the chase, he'd been a member of the force for only about five years—a period during which the department's shortcomings had been subjected to intense public scrutiny, including the Brighton 13 police brutality trial.

Daley stood nearly six feet tall, but seemed smaller, given his slight build. He took a seat in one of the four chairs at the table in the center of the room. A tape recorder was on the table. Daley appeared uncomfortable. The truth was he'd been struggling. In the weeks since the beating, Daley had sought guidance—quietly and carefully. He turned to another officer from the Roxbury police station, Jimmy Rattigan. Rattigan, the driver of the cruiser that crashed to avoid a head-on with the gold Lexus, was a union representative. Rattigan said, "He approached me; I wanna say three or four, five different times. He even called me at home; he was very upset." Rattigan liked Daley. "He was a pretty nice guy, never a problem officer or anything, a good officer." The pressing question Daley had for him: What if—what if someone saw something?

Rattigan's gut told him Daley was not one of Mike's assailants, but had information about the beating. "I felt he probably, maybe, observed something." It was a hunch based in part on Daley's anguish. "He was really bothered by this, and really worried," Rattigan said. "If he was any more upset he would have been bawling." If Rattigan was correct, Daley was caught in a no-man's-land where no cop wanted to be: torn between telling all and fulfilling his duty as a police officer sworn to uphold the law, and telling little and ful-filling his duty to the unwritten police code of silence. "He wanted to tell me, you could see it," Rattigan said. "It was in his face." But Daley never went beyond the hypothetical with Rattigan, and Rattigan told him he needed a lawyer.

The two investigators Hussey and Cruz turned on the tape recorder at 9:20 A.M. Daley began at the beginning of the night, explaining he rode alone that shift in a marked police cruiser known as the Bravo 431 unit from the B–2 station. He described his involvement in the chase for the gold Lexus, and he used a diagram of the dead end at Woodruff Way to indicate where cars stopped. His cruiser was right behind the Lexus.

Daley said he ran to the left side of the cul-de-sac, then to the right, then back and forth. While running up and down the fence, he said, that was "when I saw the person bleeding." He didn't recognize the man. "I said, 'Who are you? Who are you?'" The man did not answer. When the man unzipped his jacket, Daley saw that the man was a cop. He didn't know the cop and afterward learned his name was Mike Cox.

Daley was pretty much finished. The investigators warned Daley, "If it's proven you are being untruthful you can be terminated."

Hussey, for one, was incredulous that Daley had not seen Mike Cox when Mike began chasing Smut Brown toward the fence. Daley had come to a stop right behind the Lexus and Mike's cruiser. "You never saw anybody run right in front of you?" Hussey asked. "You didn't see him run right in front of you, is that what you are telling me?"

"Yeah. I can't see everything."

"Right in front of you."

"I don't know. I said, I don't know."

Daley's answers grew increasingly halting.

"You are not appearing very sure of yourself," Hussey said.

More than a half hour had passed, and Hussey had heard enough. "What if I told you that Michael Cox described a black male, approximately five-nine—and that someone was going to put handcuffs on him. Who do you think that would be?"

Daley said nothing. Hussey followed with another question: "Did you ever tell Officer Cox to put his hands behind his back, who you thought was a suspect at the time. Then you told him, 'Put your hands on the car.'"

Daley still said nothing. Hussey kept going; he had the floor now. "Michael Cox might have been unconscious that night but his recollection is a lot better today."

Daley did not say a word.

"Officer Daley, I'll ask you, please, don't make yourself more trouble than you have already. Okay. Be truthful with us. Don't be untruthful. It will ruin your reputation the rest of your career."

Daley spoke, not to answer a question, but to ask one. "Can I talk to my lawyer?"

Sergeant Detective Cruz shut off the tape recorder. Following a short break, Cruz fiddled with the machine. "Okay, it's 10:35 A.M. I'm turning the tape back on." The tape recorder ran for one minute more—just long enough for Daley to assert his Fifth Amendment right against self-incrimination. "I no longer want to speak to anyone."

The interview was over. Daley had made his choice. His words were the last he would ever say to Boston police investigators regarding the beating of Michael Cox.

Between February and March, investigators for Internal Affairs interviewed fifteen Boston police officers. Mostly they were stiff-armed—as when Daley "lawyered up" and shut down his interview. Jimmy Burgio's interview never got started; he showed up just long enough to assert his privilege against self-incrimination. Dave Williams, saying he had "nothing to hide," actually met twice with investigators. He then began with the canard that he and Burgio barreled into the cul-de-sac in separate cruisers. He even penciled in a phony location for Burgio's car right behind his own cruiser on a diagram. Hussey and Cruz already suspected the story about two cruisers was bogus. Hussey warned Williams about telling the truth. "I'll ask you again. Officer Burgio—was he in the car with you?" Williams admitted he was—he and Burgio did ride together.

Williams was caught in the lie. Hussey pounced. "See, David, what this is, it complicates matters if you are not being up front, truthful with us."

"I've been truthful with you the whole time," Williams insisted.

Hussey wasn't impressed. "I have a problem here," he said. "We have an officer that was severely beaten and we are pretty convinced that an officer did it. Probably mistakenly. Okay. But I have two major problems. First, the amount of beating the suspect took, who turned out to be a police officer. And secondly, when the people found out he was a police officer they walked away and left him bleeding on the ground."

Hussey was looking to leverage Williams's admission into something bigger. But Williams did not budge. He stuck to the story of complete innocence he outlined in his written report—that he'd jumped from his cruiser and caught one of the suspects after a foot chase in the front of the gold Lexus. Williams was talking about a suspect who, in fact, did not need to be chased, who was already down and accounted for soon after Williams's cruiser hit him. It didn't matter. Williams said he didn't see a beating.

Not every interview was as unproductive as Williams's. Investigators did pick up bits and pieces. In addition to Ian Daley's comments about what plainclothes officers should wear, for example, Donald Caisey added that while writing police reports Daley told him he was sure cops had beaten Mike. Investigators learned about the similar statement Dave Williams blurted out in Mike's bay in the hospital emergency room—that he thought cops had beaten Mike. Craig Jones added the information about his encounter with Dave Williams in the upstairs hallway of the Roxbury police station. Craig said, "His exact words to me, he thinks his partner may have hit Mike by accident."

Richie Walker had a tantalizing tidbit. The dreadlocked cop

disclosed he saw a Boston police officer following Mike Cox as Mike ran toward the fence after a suspect. He said the officer had to be from one of the three police vehicles that arrived ahead of him—either Cox's or Daley's or Williams's—because, he said, "no one passed by me."

But Walker's comment was a tease. His memory turned all fuzzy after that. He said he could not recall if the officer was in uniform or dressed in plainclothes, whether the officer was white or black. "I couldn't say," he said. "All I know is I saw figures going forward." Most frustrating, he denied seeing Daley, Williams, or Burgio at all.

Investigators at one point were hoping gang unit partners Joe Teahan and Gary Ryan might be able to build on these leads. After all, the pair said their car was the fourth one in. Hussey and others interviewed them three times—more than any other officers—but they got little back. Hussey said flat-out at one point he thought they knew more. But the two gang unit cops were dug in: They saw nothing. By their account, they arrived after the beating and in time to find Mike on the ground. It didn't matter that on key points they contradicted each other—or others contradicted them.

The sessions proved a disappointment to Hussey, featuring the same monotonous drumbeat of "yes," "no," and "I don't recall." After the stack of skeletal reports and a bunch of know-nothing taped interviews, Kenny Conley and Bobby Dwan came across as breaths of fresh air. The Form 26 report Kenny prepared was a detailed, typewritten account of the night that at 520 words stood out as an opus—up to ten times longer than most of the reports turned in by nearly sixty police officers. In his, Kenny wrote he thought they were the third police car at the dead end, while Bobby said they were the fourth or fifth. Both were mistaken; they were farther back than that, more likely in the seventh or eighth position. But that missed the point. The point was their openness about placing themselves in the thick of it.

When asked by the investigators to identify other officers they'd seen, Kenny and Bobby did so—naming names or describ-

ing physically those whose names they didn't know. When asked
by investigators to scan through a book of officers' faces, they did
so—pointing out the officers they hadn't known by name. In his
interview, Bobby described seeing a "commotion" down to the
right along the fence when he first jumped out of the cruiser and
ran to the left, after an officer running in that direction. When
asked about the officers down by the fence, Bobby said in that split
second he hadn't recognized them. But he did have this detail: Two
were in uniform—a black cop and a white cop.

In his interview, meanwhile, Kenny was similarly forthcoming.
While most officers seemed to take great pains to place themselves
as far away as possible from the fence, Kenny talked in the only
manner he knew—straightforward. He gave a point-by-point account of his role in the car chase and described in detail how he
bolted from the cruiser, scaled the fence, and eventually captured
Robert "Smut" Brown.

But for all their cooperation, the problem was Kenny and Bobby
didn't have any evidence to break the case. Bobby certainly had a
tidbit about seeing two uniformed cops in the commotion, but
Kenny didn't even see a commotion. He told Hussey he was coming
to a stop when down the hill he first spotted the four suspects jumping from the Lexus. "My eyes were just trained on a kid coming
out of the passenger side, a black kid with a brown leather jacket,"
he said during his first interview. Kenny ran after Smut Brown. "I
didn't see anything or hear anything," he said. "I was trained on
him." Hussey asked him if he'd seen anyone else chasing the suspect, and Kenny said no.

"There could have been," he said, "but I just kept my eyes aimed
on him."

When the sixty-five-minute interview ended, Hussey thanked
Kenny for his effort to recall the night's events. "I appreciate your
candor," he said to Kenny.

Given that, Kenny was surprised when he was called back for a
second interview several weeks later. He showed up at headquarters
on April 25 at the end of an overnight shift. The interview began at

6:50 A.M. Right away, Kenny noticed Hussey and another investigator had adopted an almost abrasive posture.

Hussey's mood swing in part reflected his frustration. The veteran cop kept expecting someone to step up and do the right thing. But no one had. He was going back to people in a position to see something, but kept getting the same evasive bobbing and weaving. He wanted to try Kenny Conley again—the rare interviewee who'd talked candidly. Intuitively, it would seem Kenny saw Mike and the assault. It was just common sense. Hussey said as much to explain Kenny's callback. "You were in really a great position here to see what went on right at that fence because that is the location, where that guy hopped the fence, that is the location where Michael Cox received his injuries."

Kenny understood Hussey's thinking. But he hadn't seen the beating.

"Do you remember seeing a commotion?" Hussey asked.

It was as if Hussey was pleading: If not the beating, a commotion?

"Out of your peripheral vision?" Hussey asked. "I know you stated before that you were focused on that suspect that hopped the fence. Did you see anybody out of your peripheral vision anywhere near the fence?"

"No, sir."

The session ended on a testy note. It was clear to Kenny that Hussey's view had changed. It seemed Hussey no longer believed him. Kenny left headquarters feeling troubled. He had certainly wondered why he had not seen anything besides Smut Brown. The question would haunt him for years to come. Kenny was like most people—like Hussey, even—who figured people see things they're supposed to see, particularly when the person is a trained police officer. Most people would guess they'd notice a beating, even while in hot pursuit of another person. But what Kenny didn't realize was this long-held assumption was plain wrong, and that scientific research conducted throughout the 1990s was debunking the popular wisdom about what people "see." Psychologists had several names for the phenomenon, "change blindness" and "inattentional blindness." Tests showed that people focusing on one event were surprisingly inattentive to something else in their field of vision that was salient and unexpected. But the research was far beyond Kenny's frame of reference. All he knew was he had not seen the beating.

Kenny drove to Southie and headed to the basement bedroom he'd built in the house on H Street when he moved back during the summer to save up some money. He needed to get some sleep. The way the interview had gone nagged at him. "I felt at that point I was being blamed for something I had nothing to do with," he said later.

The follow-up interview with Kenny Conley marked the end of the line for Jim Hussey's Internal Affairs inquiry. Ralph C. Martin II, the district attorney for Suffolk County—and the first black district attorney in Massachusetts history—was taking over the case. The switch signaled more than the district attorney's interest in the beating. It indicated Police Commissioner Paul Evans and his command staff were coming around to the gravity of the beating allegations—what one high-ranking police official began calling "among the most serious investigated by the department." The stakes were much higher than in an Internal Affairs inquiry where administrative sanctions ranged from reprimand to termination. With a criminal probe, officers faced possible jail time. But while the change of course suggested a tougher stance, it also represented the loss of even more time—a recurring theme ever since the night of the beating. Launching a criminal investigation basically meant starting over. Martin's prosecutors and officers from the police department's Anti-Corruption Unit would not have the benefit of Jim Hussey's work. They would never read about Dave Williams getting caught in a lie during Williams's interview with Hussey and Cruz. They would never read what Ian Daley said, or Richie Walker, Joe Teahan, Gary Ryan, Craig Jones, Kenny Conley, and Bobby Dwan—or Mike Cox for that matter. Being part of an administrative inquiry, evidence developed by Internal Affairs could not, for constitutional reasons, be turned over to a criminal investigation. The new investigators got written reports about the high-speed chase and dead end, but not the tapes of interviews conducted by Internal Affairs.

But if any prosecutor was up for the challenge, it was Ralph Martin. He was a young, energetic assistant United States attorney in his early forties when first appointed to become the local district attorney by the Republican governor William Weld in 1992. Then, in the 1994 fall elections, he won voters' support to a four-year term. The outcome surprised many; Martin campaigned as a Republican in a county, comprising the cities of Boston, Chelsea, Revere, and Winthrop, where Democrats outnumbered Republicans six to one. For Martin, the combination of race and justice had long been a powerful source of motivation. Originally from New York City, he was the son of a police officer. His mother was found beaten to death when he was two years old. Martin had gone to college at Brandeis University outside Boston and then moved into the city to study law at Northeastern University. He would always remember how he'd been influenced by a black assistant district attorney from Boston when he was a college senior wondering what to do with himself. The man urged a career in criminal justice because "it's important to have black folks who are principled in law enforcement."

Martin, in his brief professional life, had already displayed a willingness to go after cops suspected of wrongdoing—even if the efforts blew up in his face. Before becoming the district attorney, Martin was the federal prosecutor assigned to investigate the Boston police mishandling of the most notorious murder case in decades—the 1989 killings of pregnant Carol Stuart and her baby, shot at a traffic intersection in Roxbury after leaving birthing class. The stomach-turning case that made national headlines became more grotesque when it turned out Carol's husband, Charles, was the killer. Boston police, meanwhile, had spent weeks building a misguided case against a black man. When the shocking truth came out, the city, especially the black neighborhoods, was in an uproar, and the U.S. attorney put Martin in charge of looking into claims police framed their suspect and strong-armed blacks to incriminate him. Following his investigation, Martin recommended that "several Boston police officers should be indicted on charges of intimidating witnesses, planting evidence and violating the civil rights" of the suspect. Many officers were incensed. But their anger at Martin turned into triumph when Martin's boss, judging he could not win convictions before a jury, rejected the recommendation and decided not to prosecute the police.

It was a public rebuke, a setback fueling tension between Martin and Boston police officers that carried over into his tenure as the Suffolk County district attorney. "Conflict with the police has been a major theme in the career of Ralph Martin," noted a *Boston Globe* profile. His pro-active role in the Stuart probe "guaranteed he would get a hostile reception from many police officers when he became DA."

Now in early 1995, while monitoring the Cox beating investigation, Martin became embroiled in another police mess. He had indicted a veteran officer with forcing a prostitute to have sex while working a paid detail in downtown Boston. But the case barely got past "go." On February 27, a Superior Court judge not only threw out the rape charge, he went after Martin, characterizing the district attorney in his five-page order as an out-of-control cop hunter. The judge said Martin had manipulated the grand jury process to indict the officer, tactics he decried as a "perversion of our entire system of justice." The accused cop's attorney also had little good to say about the Suffolk County district attorney. "Martin has treated police officers unfairly," he said.

Martin considered appealing the judge's order, but three weeks later, on March 22, he decided "*no mas*"—he would not do that. He remained unapologetic, however. "Who guards the guards?" he asked rhetorically. The only cops he didn't like, he said another time, were the ones who broke the law. "People in law enforcement—not just cops—should be held to a higher standard of honesty and integrity, in order for the public to have confidence in us."

Martin said dropping the rape case should not be taken as a sign he'd lost the will to tackle police misconduct. "I don't think I have ever been shy or reluctant to investigate police misconduct when it appears," he told reporters.

It was the end of April when Jim Hussey got the order to suspend the Internal Affairs investigation and make way for Martin's criminal inquiry. Hussey walked away from the case believing he and his colleagues had made some progress, zeroing in on Jimmy Burgio, Dave Williams, and Ian Daley as the "pool of suspects." But the leads were circumstantial. There was much more to do. The IA's focus had been narrow: on the first to arrive at the dead end. Fifteen officers were interviewed, when another forty-five were known to have some connection to the chase. IA had not tried interviewing civilians living on Woodruff Way, and it had not interviewed officers from other police agencies who showed up at the scene. It had not tried interviewing the foursome in the gold Lexus who, quite possibly, had seen everything: Smut, Tiny, Marquis, and Boogie-Down.

Smut Brown and the others had been behind bars ever since their arrest, charged with first-degree murder. During the several months that Internal Affairs looked into the beating, Smut met several times with a public defender assigned to represent him. He was held in the Nashua Street jail, a new, $54 million facility built in the shadow of historic and highbrow Beacon Hill and near Massachusetts General Hospital. The public defender found him "very scared, very worried." The lawyer tried to reassure Smut, "to make sure he understood he's got help and not to lose hope." The two mostly went over the shooting at Walaikum's that started it all, and Smut was freaking out he'd been accused of murder when he'd been as shocked as everyone at the hamburger joint when gunfire broke out. Smut talked about the car chase to Woodruff Way. He said that after scaling the fence he could have outrun the cop "because the cop was thirty yards away when he called out to stop." He stopped when he could have kept going and escaped in the woods. It was the kind of detail the attorney noted. Fleeing was typically seen as "consciousness of guilt." But when Smut stopped instead of escaping, that was something the attorney could argue showed consciousness of innocence. Smut stopped because he had nothing to hide.

In these early meetings, Smut also mentioned the beating at the fence. Smut had continued to think Marquis was the one who had taken the blows. While they were being booked later at the Roxbury police station, he had not had a chance to talk to Marquis. The only thing he knew, Marquis was taken to a hospital for treatment. Smut then found out that was all wrong—from rumors, from his mother and Indira during their visits, from Marquis himself when they both were eventually in the same jail. "That was a *cop*!" Smut got to thinking. It seemed surreal. Cops beat another cop like that? This, Smut knew, was heavy. When winter turned to spring and he heard talk about the inability to solve the beating, Smut sat in jail awaiting trial for murder knowing he'd seen plenty at the fence.

"They took affidavits and reports from everyone who was on duty that night but nobody has talked to our clients," one of the defense attorneys said about the police investigation into Mike Cox's beating. The murder suspects might know something, the attorney noted, but no police official had reached out to them. "No one has contacted me to say, 'Would you mind if I talk to [them] about this other incident?'"

Ralph Martin and his immediate circle of advisers weren't the only ones in the district attorney's office who'd taken notice of the bizarre beating case during February and March. Assistant District Attorney Bob Peabody had read the first newspaper account. Peabody, thirty-nine, worked in Martin's Special Prosecutions Unit. His bloodlines ran deep in Massachusetts history. He was a descendant of the founder of the Massachusetts Bay Colony and the state's first Colonial governor—John Endicott. His father, Endicott "Chub" Peabody, was governor in the early 1960s. Peabody, tall and built solidly like his father, played the line on the Harvard College football team. He studied law at Boston College. With about seven years of experience as a prosecutor, Peabody was no longer green but had yet to attain the status of seasoned veteran.

Bob Peabody noticed the story because he knew Mike Cox. More accurately, he knew who Mike Cox was. Peabody's kids attended the Park School in Brookline with Mike's two boys. Peabody's middle son and Nick Cox were in the same elementary school class. He and Mike traded greetings at school events, but little more. When Peabody read about Mike's injuries, he made a point of going up to Mike the next time he saw him at the boys' school to ask if he was doing okay. Mike was polite and friendly enough, but it was a brief, awkward exchange, the kind Mike was trying his best to avoid. Soon enough, however, Bob Peabody and Mike Cox would be having more formal exchanges about the beating.

For Boston Red Sox fans, springtime was always about a fresh start, a new season of hope for winning a World Series and ending the miserable drought since the last national championship in 1918. The baseball strike in 1995 that was threatening to shut down the Major League Baseball season had ended on April 2, and three weeks later the fans piled into Fenway Park for the home opener. The worry was that hard-throwing ace Roger Clemens was out of action with a shoulder strain. The good news was the lineup: Slugger Jose Canseco had been added to provide more home run power, while shortstop John Valentin was maturing into an elite player. Down in the minor leagues, a new infielder named Nomar Garciaparra was impressing with his slick fielding and batting stroke. The Sox took the field and clicked on all cylinders: Pitcher Aaron Sele threw a shutout and Sox hitters drove in a slew of runs in a 9–0 win over the Minnesota Twins.

For Sergeant Mike Cox, the opposite was true about the spring: It was a season of despair and hopelessness. "My wife was saying, 'I told you this wasn't going to go away. This is terrible what they did to you. Don't you see it?'" By the time Jim Hussey was shutting down the IA inquiry, Mike had indeed begun to come around.

"I mean nothing's happening. Nothing's being resolved."

With help from Craig Jones, Donald Caisey, and a few other friends, Mike monitored the department's faltering Internal Affairs inquiry. He picked up bits and pieces. He learned Dave Williams owned up to lying that he and Jimmy Burgio rode in separate cruisers, but he also learned nothing happened to Williams for screwing up the initial effort to establish the layout of cops and cruisers at the dead end. Williams had a knack for eluding trouble. Just as he was becoming embroiled in the unfolding Cox affair, he appeared to be facing hot water on another front. Two civilian complaints against him for excessive force, still pending from the fall, caught up to him. The cases, at long last, had triggered the department's new Early Intervention System, the program to monitor officers with two or more complaints. It had taken four months—and only after the Cox beating probe began—when Williams was notified to undergo retraining due to a "potential problem area: physical abuse during arrests." He was ordered to enroll in a course in "Verbal Judo" while the complaints were under investigation. Verbal Judo was a method for using words rather than force to defuse volatile situations. But Williams quickly learned EIS was more sound than fury. The city thought Verbal Judo was not a proven program and wouldn't pay for it. Williams never took it or any other course. Instead, he focused his verbal skills on denying the excessive force accusations still under investigation, including the one by Valdir Fernandes, the eighteen-year-old who accused Williams of smacking him around for spitting too close to Williams on a porch.

Mike also learned about Ian Daley taking the Fifth. It made him mad that someone he believed definitely had evidence about the beating had gone stone cold. But if Mike was upset, Craig Jones was furious. Craig was convinced Daley was either a beater or a witness to it. "I was pissed," he said. "I couldn't believe no one was going to step up." He was also taken aback that many cops didn't seem to realize the severity of the beating. "I had people coming up to me saying something like this happened to them. I'd say, Oh, you did? You had blood in your urine?" Craig tried to set them straight. This was not a case of one cop mistakenly spraying another cop with mace or whacking him with a glancing blow, "and the cop immediately apologizes and it's over." This was much worse. "It was a crime," he said. Mike was owed "a lot more than an apology.

"Some things you can let go, but some things you can't, and this was one of those things," he said. "If you were pissing blood, I'd tell those guys, would you let it go?"

Throughout the three months Internal Affairs was investigating the beating, Mike continued to puzzle over Police Commissioner Paul Evans. The commissioner was less aggressive than he'd expected. "Initially he didn't take it seriously enough," Mike decided. The top cop's generally silent posture, he thought, had created a vacuum where the downplaying of the beating that Craig Jones kept running into was able to gain traction. It was the fertile ground for rumors questioning the department's commitment to hold the beaters accountable. Was Evans serious? Or should everyone sit tight and ride out a whitewash? One thing was certain. In his first year, the commissioner was proving himself as a master of policy statements. He issued a series of impressive and high-minded new directives aimed at improving the department's tarnished image.

Less than two months after the beating, Evans in March promulgated a new order explicitly listing the duties of sergeants and patrol supervisors when force has been used by officers in their command. It amounted to a checklist of what the three sergeants at Woodruff Way—Isaac Thomas, David Murphy, and Daniel Dovidio— did *not* do following Mike's injuries. The order was called "Special Order Number 95–16: Amendment to Rule 304, Use of Non-Lethal Force." Its purpose was "to more clearly delineate what constitutes a full and complete Patrol Supervisor's investigation in cases where the use of non-lethal force is used, or alleged to have been used, on a subject." The new rule required that "prior to the end of the tour of duty," the supervisor was to prepare a report that included reports from officers alleged to have used non-lethal force, reports from all police personnel at the scene, and reports from civilian witnesses.

Two days later, on March 17, Evans announced a second new directive called "Special Order Number 95–17: Identification of Plainclothes Officers." Its purpose was "to minimize the potential risk to officers assigned to plainclothes duties by establishing policies and procedures that will aid in their being properly identified." The rule, which had been in the works for months, established for the first time a hand signal officers working in plainclothes could employ to identify themselves to other officers—"in order to avert an unfortunate confrontation or tragedy." The officer was to raise his arms over his head and cross them at the wrists, turn his palms forward, and spread out his fingers.

Then several months later came Evans's new "Public Integrity Policy: Rule 113." The nine-page, single-spaced directive addressed the need for the police department to "maintain the highest standards of honesty and integrity." It noted that police departments throughout the country experience corruption. "Boston certainly has not been immune to those problems. Corruption, brutality, falsifying evidence, and bias cannot be tolerated among individuals sworn to uphold the law. Nor can hypocrisy, unfairness, deceit and discrimination be tolerated in an organization dedicated to the highest ideals of justice and the rule of law."

The document contained eleven ethical canons. Canon nine, in particular, stipulated: "Police officers shall use only that amount of force reasonably necessary to achieve their lawful purpose. Excessive or unauthorized force is never justified and every officer not only has an affirmative duty to intervene to prevent such violence, but also to report any such instances that may come to their attention."

The orders addressed flaws exposed in dramatic relief by the beating. To Mike, the words certainly looked good on paper, but what about their application? What about the lies and failure to cooperate during the Internal Affairs inquiry? Truth telling could have been a concern pursued in any number of directions—Dave Williams and Ian Daley, to name two. No one was looking at Sergeant Dan Dovidio. Even cohorts in Mike's gang unit were vulnerable, given the fantasy about the ice-slip codified in reports that first night.

Beyond the lying, there were leads that seemed to be ignored. What about Richie Walker saying he saw a cop chasing Mike; Bobby Dwan saying he noticed a commotion by the fence involving a black cop and a white cop in uniform; Mike saying he was kicked by a white officer; gang unit officers saying Dave Williams and Ian Daley later said cops had beaten Mike; Craig Jones saying Dave Williams later had told him his partner had hit Mike; Ian Daley dramatically shutting down his interview, and Jimmy Burgio taking the Fifth? The burden of proof in an administrative inquiry was substantially less than the "beyond a reasonable doubt" required in criminal cases. Why were Burgio, Williams, and Daley still on the street? Mike wondered why they weren't at least put on desk duty while the investigation into the beating continued.

"No one's getting in trouble," Mike said. The commissioner's directives seemed like big talk, little action. The message cops could take from the absence of any real consequence was to hunker down: No one else is saying anything. I'm not saying anything.

"I was feeling more and more uncomfortable with the process," Mike said.

By early spring, it didn't take much to awaken Mike. He slept on edge, as if waiting for something to happen. It could be a telephone call; the crank calls had continued as February passed into March and March into April. Or it could be the nightmare in which he was helpless against police attacking his house and family. Or it could be something else. Mike was always wondering, what next?

The pounding at the front door therefore saw Mike bounding out of bed and hustling downstairs in his shorts and T-shirt. It was the middle of the night, and he didn't want his boys or his sisters' family downstairs waking up. Kimberly, however, was out of bed and right behind him.

Before Mike reached the landing he heard the loud crackle of police radios outside. He opened the door. Two uniformed officers stood on the stoop. Mike recognized one of them, but didn't know his name. They were from the B–2 station in Roxbury.

The second officer, the one Mike did not recognize, spoke up. He said they'd been dispatched to the house on a 911 emergency call—a 911 call, the officer said, for a disturbance. "For a man being beaten."

Man being beaten? Mike couldn't believe what he was hearing. You gotta be freakin' kidding me, he thought. *Man being beaten?* The beating was January 25. Woodruff Way.

The two officers seemed poised, ready to barrel into Mike's house.

Mike didn't say a word. He filled the door frame. It was a silent standoff lasting a few seconds. The cop Mike had recognized then recognized him, and he turned to his partner.

"Let's go," the cop said. The second officer, confused, hesitated. The first kept going, heading toward the cruiser parked on the street. C'mon, he called back to his partner. To Mike he said, "Sorry."

Mike shut the door firmly. Kimberly wanted to know what was going on. "Just go back in," Mike said. Upstairs, he told her they had the wrong house. It was a mistake. Oh, was all Kimberly said. They got back into bed and tried to get some sleep.

Both knew Mike was lying. It was not a mistake but a new twist in the ongoing harassment—a phony 911 call, almost funny in a perverse way.

Mike understood that inside the department, there was no longer any question about where he stood. Everyone knew he was cooperating. He'd met with Internal Affairs and made it clear he wanted justice.

With that, it seemed to Mike the message behind the harassment was changing. It had gone from being a warning to lie low and not cooperate to a kind of punishment and payback for deciding to push the matter. The new message was: You're not one of us.

Thoughts like that were taking their toll. Instead of feeling better three months after the beating, Mike was feeling worse. It was blowing his mind—he'd been beaten, he was the victim, and yet he was the outcast.

"What is it about me?"

He found himself obsessed with the question. "What is it about me that these people think that they can just do this? And just walk away, and never admit to anything or apologize."

He was thinking in ways he never had before. He was never particularly race-conscious growing up, but he began thinking that being black was the only viable explanation for the abandonment that started at Woodruff Way. "They were able to leave me because they thought less of me because of what I am," Mike said. "It wouldn't have happened if I were white."

The thoughts at times had a crippling effect. Mike retreated into himself. He became jumpy and fearful, and, on occasion, Kimberly found him alone, crying. But something else was at work too; at other times the thoughts worked Mike into a rage. "The more I relived what happened to me, the angrier I got," he said, "and it wasn't just the beating that angered me. It was the fact that they left . . . They just left me there like an animal to die, you know, on the side of the highway."

While the Internal Affairs inquiry was winding down, Mike made a key decision. He contacted the Boston attorney who'd come to the house to talk about possible legal action. During the first visit, Mike hadn't given Stephen Roach the time of day. But Mike now told Roach to go ahead—Roach could write the city to request a full copy of Mike's medical records. It wasn't as if Mike had committed to taking specific action, but he no longer knew whom to trust and he wanted to be ready. He saw no harm in letting Roach make a few preliminary requests for documents.

Mike also agreed to a second request from Roach. He agreed to meet with a psychiatrist. He didn't believe in therapy, but the lawyer wanted a professional evaluation. Seeing a psychiatrist might also ease some of Kimberly's concerns. Mike went twice to see an analyst named Dr. Jerome Rogoff—for ninety minutes on April 24 and for another sixty minutes on May 1. Rogoff was a forensic psychiatrist with a drawer full of credentials. He was a member of the American Academy of Psychiatry and Law and had taught at Tufts

Medical School and Harvard Medical School. He'd treated prison inmates and served as an expert witness in both civil and criminal cases.

Following the two sessions, Rogoff wrote Roach a report that ran six pages. It began with an outline of Mike's life, saying that Mike grew up in Roxbury in a "stable, intact family." He noted that his father died of cancer when he was sixteen, "at which point his severe and prolonged bereavement interfered with his performance at school." By senior year at Wooster, the report said, he'd recovered and finished strong.

The main thrust of the report, however, was not Mike's past. It was to determine Mike's "current mental and emotional state." Rogoff found Mike "guarded and suspicious," which, noted the psychiatrist, was not typical for him. "He was always preternaturally calm under stress, which contributed to his effectiveness as a plainclothes policeman." The psychiatrist listed other symptoms all too familiar to Mike and his family—"he has trouble sleeping," "the content of the nightmares is of people trying to shoot him, to kill him, often while he is with his children," and "he cries from time to time, especially when he thinks about how even those he thought of as his friends on the police force have deserted him."

The diagnostics ran several pages. "Taken together, these symptoms describe a Post Traumatic Stress Disorder with elements of clinical depression (Adjustment Disorder with Depressed Mood) added to it."

Rogoff was clear about the cause of Mike's troubles. "Both the timing and the content of these symptoms, coupled with the fact none of them existed before January 25, 1995, leave no question that they were caused directly by the incident of that night."

It was discouraging stuff. And Rogoff wrote there was no specific treatment for post-traumatic stress disorder. But he did end with a note of encouragement. "I was impressed with Michael Cox's inner strength and integrity of character (personality), and thus I would hope that he would be able to put himself back together psychologically in a fairly short time."

CHAPTER 12

Dave, I Know You Know Something

Two years before the beating of Mike Cox, a pair of leading experts on police practices were asking: "How can police, who can be exemplary heroes, beat people and then even be prepared to lie about it?" The paradox was the central question explored in their 1993 book, *Above the Law: Police and the Excessive Use of Force*. The two scholars were Jerome H. Skolnick, a law professor at the University of California, Berkeley, and later New York University, and James J. Fyfe, a longtime New York City patrolman who left the force after earning his Ph.D. in criminal justice.

Using cases that included the ferocious beating of Rodney King on March 3, 1991, by Los Angeles police officers—captured on videotape and televised around the world—the authors said the answer was found in the proposition that "two principal features of the police role—danger and authority—combine to produce . . . a distinctive world view." It's an "us-versus-them" perspective, where the high-risk and often violent nature of the job creates a policing culture based on "internal solidarity, or brotherhood."

The brotherhood, they wrote, controls behavior even when an officer crosses the line—such as in the beating of Rodney King. It almost doesn't matter that police departments routinely issue policies on integrity and truth telling. (In Boston, it was Commis-

sioner Paul Evans's new "Public Integrity Policy: Rule 113.") When
it comes to survival on the street, the unwritten codes about sticking together are what matter, even if that means lying and a cover-
up. The last thing a cop wants to do is testify against another cop.
"The code decrees that cops protect other cops, no matter what,
and that cops of high rank back up working street cops—no matter
what," wrote Skolnick and Fyfe.

Mike Cox did not need to read any book to understand the blue
wall of silence. "It's a large part of being a police officer in general
and the culture of being a police officer—protecting one another."
In his analysis, the code's logic began with the presumption that
a cop was always right. "Whatever it is that he's doing is assumed
to be right. Because you're assuming his actions are always right,
you don't look for any wrong." In other words, there was never any
misconduct for cops to talk about. It was why one of the officers
on the witness stand in the Brighton 13 police brutality trial could
unblinkingly testify he'd never seen a Boston cop doing anything
wrong. In his six years on the force, Mike certainly understood the
basis for the code—the need to be able to count on your cohorts,
virtually without reservation. It was all about survival while working on the edge of life and death to uphold law and order.

But then came his beating. The code's underlying rationale—
"us versus them"—did not fit. Mike was not one of "them." He was
not a drug dealer. He was not a gun-toting gangbanger. He was not
Rodney King. Mike was a cop—one of legions of "us" that made up
the Boston Police Department. If the beating had been posed to
him as a hypothetical scenario, Mike would have said that the code
did not apply—not when the crime victim was another cop. But that
hadn't happened. Instead, Mike watched good cops—good guys,
friends of his—not wanting to get involved because doing so meant
testifying against another cop. This included black cops—when
he would have thought a racial solidarity among the black officers
would have kicked in to override the silence. But that hadn't happened either. The code had a power Mike never imagined. It even
trumped race. While a second investigation began in the ashes of

the first, Mike was seeing that the reach of the blue wall extended beyond concern for any one individual. "And I just happened to be the individual."

In June, the *Boston Herald* reported that District Attorney Ralph Martin and the police department's Anti-Corruption Unit had taken over the investigation of Mike Cox's beating. "This was nothing but another Rodney King situation—only this time there was no video camera," one of the paper's unnamed police sources was quoted saying. Since it was located inside the tabloid, the casual reader might easily have flipped past the story. But interested parties—meaning the police world—would certainly notice, and for those readers there was more: The story named names. "Two officers from Area C–11," it said, "James Burgio and David Williams, have been questioned in connection with Cox's beating, but no formal charges have been brought against them, sources said."

The account was brief, only 408 words long—hardly headline coverage. The rest of the Boston media barely blinked. In fact, the report was only the fourth article since the January 25 beating. Such a low story count was surprising. For one thing, the chase was considered the longest in Boston anyone could remember, involving the most cruisers ever. Then a high-ranking police official began referring to the assault as "singular" in the department's history. Police Commissioner Evans's chief of internal investigations, Ann Marie Doherty, based her characterization on the combination of "the type of injuries that were sustained and the fact that medical attention was not immediately provided." The newsworthiness of the case was crystal clear. It was a no-brainer: Journalism 101.

But the story hadn't gained any traction. It wasn't as if the local press was not busy covering the police department—good news and bad. Commissioner Evans and Mayor Tom Menino, along with a number of city officials, posed for photographs at the ground breaking for a new police headquarters in Roxbury to replace the seventy-year-old building in the Back Bay that was obsolete. It would take two years to complete the new $62 million structure Evans boasted would house state-of-the-art ballistics and crime labs, and the relocation to Roxbury was viewed as improving the department's community policing programs along with its accessibility to residents.

There was the bad press Evans and the mayor continued to get over Accelyne Williams's death during a raid the year before. Negotiations with the elderly minister's widow, Mary H. Williams, had broken off, and she'd sued. One of her lawyers called the bungled drug raid that led to the minister's death by a heart attack "a crime that the city of Boston and the police department have to be punished for." Mayor Menino's press office released a statement calling the suit "regrettable." It noted the city's proposal to pay $600,000 was "the largest litigation settlement the city has ever offered." But that response only infuriated columnist Derrick Z. Jackson of the *Boston Globe*. "Rodney King," wrote Jackson, "collected $3.8 million." Mayor Menino, he wrote, was Mayor Scrooge. "Williams vomited and died of a heart attack. Williams was a pristine victim, with not a scintilla of warranted suspicion. One could not possibly imagine a more obvious case to settle quickly, not just to ease the pain of the family but also to send an expensive signal to police that they had better learn to tell black people apart."

Then, of related interest, there was the resolution to the police brutality case in nearby Providence, Rhode Island, that had been making news since the week before Mike Cox's beating. Four months after his suspension the rookie cop who'd been caught on tape kicking a black concertgoer was allowed back to work. There were strings attached to the reinstatement. The officer was required to repeat his training at the police academy and would not be allowed to work on the street until he did. He had to take a course on cultural diversity. He forfeited $12,000 in pay. Finally, he was required to write an essay. The topic: what it meant to be a police officer as it related to human rights.

But all of those were breaking news events that the media then reacted to and covered. The Cox case was different. No one was holding press conferences or handing out press releases. In fact, no one connected to the scandal wanted any coverage—not the police commissioner, not the mayor, not the police union, not even Mike Cox, given his nature. Boston did not have its own Reverend Al Sharpton, the New York Baptist minister and vitriolic activist, to inflame public interest. Instead, the Cox case belonged to that category of news story the media uncovered by being investigative, enterprising, or "pro-active" in its reporting—and in this instance the Boston media had dropped the ball.

Bob Peabody, the assistant district attorney leading the new criminal investigation, was just returning to the office, fresh from finishing a special assignment working alongside federal prosecutors in winning the racketeering convictions against a number of gangsters from the city's Charlestown neighborhood. The verdict on March 22 followed a grueling five-month-long trial, and an investigation that had taken several years. The outcome was hailed as a major break in the largely Irish neighborhood's notorious "code of silence," where residents, or "Townies," were loath to testify against one of their own for fear of retaliation from neighborhood thugs. Deeply embedded in Charlestown's insular ways, the code was a key factor behind a shockingly high unsolved murder rate. In two decades, nearly 75 percent of the fifty murders remained unsolved—far exceeding the rate in any other section of Boston. It gave credence to the local slogan: In this town you could get away with murder. But the trial verdicts now suggested otherwise. Led by a young, aggressive federal prosecutor named Paul V. Kelly, investigators convinced residents to cooperate and even persuaded some gangsters to turn against their cohorts. The government spent nearly $1 million to protect and relocate up to eighteen witnesses. "We're not going to turn around 100 years of history with one case," Kelly said afterward. "But hopefully we have dented the code."

Bob Peabody was on trial with Kelly when he'd first read about Mike's beating in the newspaper in late February. Once the trial ended in late March, he began working his way back into the district attorney's office, and during his transition one of Ralph Martin's top aides asked about his interest in taking on the Cox case.

"Yeah, absolutely," Peabody said. He liked the idea of digging deep into a new investigation, and he hoped to apply some of the tips he'd picked up during the federal investigation. "I'd had practice and experience on the federal side doing this long-term investigation in the Charlestown case, so it actually was a great chance to do it again."

In a way, he was going from one "code of silence" case to another. The two worlds were obviously different in so many ways—an entire neighborhood, on the one hand, compared to a police force. But both were tightly knit, insular, and seemingly impenetrable. Police officers here and there had offered tidbits, but the blue wall of silence was proving durable. Police Superintendent Doherty, the department's chief of internal investigations, acknowledged as much, writing at one point in court papers that Boston police officers weren't talking "because they fear retaliation, harassment, intimidation and unfavorable decisions on promotions and assignments."

Peabody chose not to dwell on comparisons between the two cases or reasons that police officers had not come clean. He viewed Mike's case as a straightforward investigation of an assault and battery. "Our job was to try to figure out how he got hit and who hit him. That was our objective." He considered instead the other obstacles—such as the nearly four-month delay in getting started. The first twenty-four hours after the beating, he knew from experience, had been the best chance "for people to figure it out and cut to the chase, and to be big about it." Once lost, "It was every man for himself."

Then the Internal Affairs investigatory materials were off-limits. "We were barred from obtaining any of the information from Internal Affairs." Peabody and his Anti-Corruption investigators would be unable to challenge any changes in the officers' tape-recorded statements—known as prior inconsistent statements. But it was more than comparing content. Interviews were about body language too, and experienced investigators studied words and body language in assessing an officer's credibility. In this regard, the IA interviews were tantamount to dress rehearsals—a practice round for testing statements and delivery. Peabody's interviews were not going to be fresh.

Despite his role in the Charlestown affair, Peabody was also the first to admit his own relative inexperience when it came to using the grand jury as an investigatory tool. Ordinarily prosecutors went before a grand jury to seek indictments based on evidence already assembled by the police. Typically, this legal step was brief and uncomplicated. In the Cox case, Peabody, as the lead prosecutor, would be trying to *develop* evidence in front of the grand jury, where he'd be calling witnesses to the stand and attempting to build a case by "probing and digging and pushing." But if he lacked the sure-footedness of his later career, when he would run a number of investigatory grand juries, Peabody entered the Cox case eager and undeterred. He attacked the case with the same determination he showed while playing tackle on the offensive line of the Harvard College football team. "We'd start from scratch, ground zero, and build our own case."

The game plan wasn't fancy or revolutionary. Peabody began by reading the stack of police reports about Woodruff Way, underlining certain statements and jotting notes in the margins. As he began to reconstruct the night, he couldn't help but get swept up in its high drama—the wildness of the high-speed chase for the four shooting suspects followed by the "fucking chaos" at the dead end. "The adrenaline was pumping like you wouldn't believe." It reminded him that life for street cops was "God-forsaken work. You're working on the edge. Your life is in your hands. It's scary. People don't get that."

Peabody was teamed up with the head of the police department's Anti-Corruption Unit, Lieutenant Detective Paul J. Farrahar. The two men knew of each other but had never worked together before. Farrahar, about to turn fifty-four, was a commanding figure. Like Peabody, he stood more than six feet tall. But Farrahar exuded a physicality that Peabody did not, despite Peabody's history as a college football jock. Peabody was a blue blood, with an Ivy League polish and hint of the Boston Brahmin in his voice. He was no snob, for sure; he was down-to-earth—he'd eaten his fair share of turf as a lineman—but his earth was different from Farrahar's. The balding cop's background was working class. His handshake was firm, driven by powerful forearms, and he possessed an unflappable demeanor. The inscrutable look, however, did not mask a hard interior. He had seen it all during his twenty-five years on the force—or thought he had until the Mike Cox beating. The more he learned, the more worked up he got. The silence that followed was beyond his comprehension. It turned his insides that cops had run from a fallen cop.

Together, Peabody and Farrahar began scheduling the interviews in the case officially known as "ACD Case 95–12: Sergeant Michael Cox, Assault and Battery." Farrahar typed up a list of nineteen questions to serve as a guide. The list began with the basics. "State your name, rank and ID number." It ended with the heart of the matter: "Did you see any person assault Michael Cox?" They decided to ask key officers to draw by hand a diagram of the cul-de-sac at Woodruff Way, marking the locations of cruisers and officers, an exercise that resulted in a collection of wildly different pictures. For their use, Farrahar prepared a rough drawing of Woodruff Way—a diagram he then taped to a piece of shirt cardboard from the cleaner's—so they could position the cruisers and people based on evidence they developed during the course of the investigation.

Unlike the Internal Affairs Division, the BPD Anti-Corruption Unit was housed "off campus," or away from police headquarters on Berkeley Street in the Back Bay. Just recently, the unit had moved into new offices in the Fort Point Channel neighborhood. For much of the 1800s and into the 1900s, the area was a vibrant shipping and industrial center, bustling with ships, warehouses, and brick and granite factories. More recently, Fort Point Channel had become an afterthought, known for its funky studios and small start-up businesses, as artists and entrepreneurs took advantage of the rundown warehouses and cheap rents. But change was afoot again. Since it bordered the financial district, developers increasingly had their eye on the neighborhood's potential. It was officially an area "in transition" when Farrahar set up shop on the fifth floor of a red-brick building on Congress Street. The offices, freshly painted and newly carpeted in a speckled gray-blue pattern, were in the building's rear. The idea behind having such a nondescript location was to provide some privacy, so that police officers summoned to meet with investigators could come and go without fanfare.

Both Peabody and Farrahar joined the investigation well aware of the simmering bad blood between the DA's office and many rank-and-file cops over Ralph Martin's track record of going after cops. Peabody wasn't going to let politics affect his work. Indeed, Martin had told him as much. "Ralph said, 'Just do it.'" But that didn't mean Peabody had no appreciation for the pressures at play. He was clear-eyed. "You've got to win these." Losing a corruption case against a cop, he said, "would be devastating."

Peabody's first interview was with Mike Cox. The session on May 11 fell a week after Mike was diagnosed by his psychiatrist with depression and post-traumatic stress syndrome. Once again, Mike shared what he remembered about Woodruff Way. Peabody, Farrahar, and a third prosecutor listened quietly, occasionally asking questions. They had him look at some photographs of Dave Williams's cruiser. Taken by a police photographer to document the cruiser's damage, the photos also showed the rear trunk streaked with Mike's blood. Mike talked about Ian Daley nearly arresting him. When they were done, Peabody was struck by how "mild mannered" Mike was. He also sensed Mike's despair. "He's an island unto himself at this point." The second prosecutor scribbled notes as Mike covered a span from when he joined the police department to his treatment in the hospital emergency room. On the last of seven pages the prosecutor highlighted two names: "Bergio," which was spelled incorrectly, and "Dave Williams."

Peabody wished Mike could have given them more. In the ideal case, Mike would have been able to tell him the identity of the beaters. "He couldn't pin the tail on the donkey," Peabody said later. "He just couldn't." But Peabody was satisfied Mike had done the best he could, taking his injuries into account, and he launched his investigation feeling confident. He would proceed methodically with the grand jury, working out from the victim and moving from car to car and from cop to cop. "You investigate . . . you explore, you probe." He was in no rush. "We had plenty of time. We were going to do this painstakingly." He was also under no illusions. The investigation was going to be a long haul. But he was hopeful about the prospect of bringing the beaters to justice, about developing enough evidence—a mix of direct and circumstantial—where the only "reasonable inference" was to convict Mike's assailants.

"That's not a bad way to go to court—it *has* to be them!"

Nearly a year later, Bob Peabody would feel otherwise.

Its newsworthiness notwithstanding, the probe of the Cox beating continued below the Boston media's radar under a cloak of darkness, talked about by police in rumor and whispers in station house locker rooms, out on patrol, or in the bars after shifts, often crackling with a tension that, on occasion, erupted.

Craig Jones, for one, had trouble containing his anger—and he targeted Ian Daley as an outlet for his frustration. "After Mike told me about Ian Daley trying to handcuff him, I'm like, 'That's got to be the one guy that knows exactly what happened, the guy standing there with the handcuffs.'" Late one night Craig happened to walk into the Roxbury station house and spotted Daley already in the front lobby. Craig sat down near the front desk to write up a report about an arrest. Daley stood not far away.

Craig looked over. "Why don't you come clean?"

Daley said he didn't know what Craig was talking about.

"Just come clean, Ian. You know what happened."

Daley again said he didn't know what Craig meant.

Craig couldn't take it. "You're a liar," he yelled.

Heads turned. Craig swore at Daley and Daley yelled back. The sergeant on duty got up and came around from behind his desk, "Hey! Hey! Knock it off!" He then ordered the two officers into a side room, where he kept them until they cooled off.

But it didn't end there. The two had a second run-in. Both responded to a late night shooting several weeks later on Lawrence Avenue in Dorchester. Craig, working in plainclothes, was walking up the street when he saw Daley walking in the other direction. Immediately Craig began gesturing theatrically—raising his arms over his head and crossing his hands at the wrists in mockery of the new identification signal implemented by the police commissioner for plainclothes officers to give to uniformed officers. "I'm a police officer!" Craig yelled melodramatically. "Don't hurt me."

"Fuck you," Daley said.

Daley complained to his supervisors at the Roxbury station, and, once again, Craig was told to knock it off. Craig got no satisfaction from the run-ins, but couldn't help himself. "When I encountered Ian Daley," he said, "I'm upset about what happened and how all the officers who were there *for some reason* never saw a damned thing."

Jimmy Burgio and Dave Williams, meanwhile, were carrying on as usual. They showed up for work and performed their regular shifts. To earn more, they worked paid details. For Burgio, the extra money was welcome. On Saturday, June 24, he married a Dorchester woman he'd met earlier in the year. Williams went, toasting Jimmy's big day, while Burgio's longtime partner, Lenny Lilly, served as one of the ushers. For Burgio and Williams, the probe was little more than background noise.

Around the time of his interview with Bob Peabody, Mike and Kimberly headed over to Franklin Park one sunny weekend to take advantage of the warmer weather. The park, the city's largest, with 527 acres, could be a dangerous place after hours. Rape and murder on the park's grounds were an unfortunate and dark side to its history. The night of the beating, Mike, Craig, and the other officers had chased the gold Lexus along the park's east side, roaring down Blue Hill Avenue. But the daytime was entirely different. The park was home to a zoo, a golf course, and playing fields for baseball and soccer. Families picnicked on weekend days. Joggers and bicyclists dotted the pathways.

The long walk was a chance for husband and wife to be alone and to get out of the crowded, densely built street where they lived. Kimberly's graduation from medical school was only weeks away, and Mike was going to travel to Philadelphia with their sons in June to attend this milestone in his wife's career. Kimberly wasn't going to be able to sit on her laurels for long, however. On July 1 she would begin a one-year internship in internal medicine at the Carney Hospital in Dorchester. It was the same hospital where Mike, Craig Jones, Richie Walker, Dave Williams, and others sometimes met for a meal in the hospital cafeteria after coming off an overnight shift.

July was also looking to be a big month for Mike. It was when he was scheduled to return to the force. Physically, he'd been coming around. He no longer wore the splint to stabilize the damaged ligament in his right thumb. With therapy, the thumb was feeling stronger. He could hold his service revolver okay. The thumb still swelled easily if he used it a lot, so he quit playing for the gang unit's basketball team in the police league. He usually began playing tennis at least once a week during the spring, but at this point he didn't give playing even a passing thought. His urine was still brownish in color, and the severe headaches dogged him. Taking Advil or Motrin was part of his daily diet. But his amnesia was wearing off, along with his dizziness and occasional disorientation. He'd not had another freaky episode like the one when he was driving home from a doctor's appointment and "I just drove by my house. I don't know where I was driving, but I had gone past my house, way past my house, and I realized, you know, Where am I going?" In five months, he'd had more than thirty visits with a dozen or more doctors.

Mike wasn't going to be rejoining the gang unit. He had a new assignment—and of all places, the newly promoted Sergeant Mike Cox was headed to the Internal Affairs Division. It seemed surreal: joining the division that had failed to solve his own beating. But the transfer had been in the pipeline well before the night of January 25. It was generally believed career-minded officers—officers who aspired to high-ranking positions on the force—needed to rotate through Internal Affairs or Anti-Corruption. With that in mind, Mike had actually sought the assignment. But now Mike no longer felt so ambitious. He no longer knew what to think about his career. He wasn't ready to quit, which was what Kimberly and others in his family wanted. But his career seemed in shambles to him. His mind was preoccupied around the clock with the case. He didn't know what to do, except to go ahead and report to work in July and see what happened.

The couple had all this and more on their minds as they were making their way through the park. Then they heard someone calling out Mike's name.

Mike looked and recognized Dave Williams.

Hey, Mike. How you doin'?

Mike was surprised to see Williams, but Franklin Park was also a familiar destination for him. He jogged in the park if he skipped his early morning workouts in the gym. Mike realized it had been a couple of months since he and Williams had talked. Kimberly strolled ahead and left the two men alone on the path.

Williams took the lead by asking a question: I hope you don't believe that stuff that they're saying in the paper? Mike didn't answer. He listened. Mike, I know you, Williams said, I know you. Williams repeated the line, or a variation of it, more than once: *You know I know you, Mike. You know I wouldn't hit you.*

The refrain had the sound of a talking point. Then he delivered a second line: He switched subjects from himself to Jimmy Burgio, and when he did Mike noticed something. "He was talking to me face to face," Mike said, "and then when he got to that part—'And as far as Burgio, ah, well, ah'—he wouldn't look at me." To Mike, Williams was not acting like the guy he'd known for almost four years: confident, direct, up tempo. "He said, 'As far as Burgio, well, I don't know how they, you know, could say that because, ah, well, he was kind of . . . You know, he was right behind me.'"

The gist of the halfhearted rambling was to back up Burgio, and Mike couldn't stomach listening to any more of it. "Dave, c'mon," he interrupted.

Williams stopped. Mike continued. "This thing is not going to go away," he said. "It's not going to end any time soon. It's only going to get worse." Mike then began his own talking point, telling Williams several times, "Just tell the truth."

When Mike finished, Williams didn't try to pick up where he'd left off. Nor did he respond to Mike's challenge by insisting he was telling the truth, as he had to Jim Hussey during the interview with Internal Affairs. Williams just didn't say anything more about that night at the dead end, and the accidental meeting between the two ended there.

Mike turned the exchange over and over in his head, looking for meaning beneath the surface of the word choices and elliptical sentence constructions, the repeated *You know I know you, Mike*, and so forth—all part of a haunting puzzle to him. It seemed more fitting for a code breaker working intelligence in the world of international espionage. By his interpretation, Mike decided the faint defense of Burgio—delivered in a stutter and humility uncharacteristic of Dave Williams—was his former friend's way of signaling Burgio was culpable but that he was bound to cover for him.

It all left Mike feeling empty.

Two months later Mike again unexpectedly ran into Williams. The occasion was the funeral in July of a fellow officer, Sergeant Diana Green. Dee Green had committed suicide. The news came as a blow. The last time Mike had seen Dee was when they'd stood proudly together on the stage during their promotions to sergeant. She was a friend, and Mike had always respected her accomplishments. When she did not show for work, her captain had gone to her condo in Roxbury. He got in and discovered her body on the floor. Her dog, a German shepherd named Buddy, stood by. There were no signs of forced entry. In the *Boston Herald*, columnist Peter Gelzinis wrote a tribute. He quoted one of Green's former supervisors. "Her courage was a given," the supervisor said. "I watched Diana make a stop of two guys in a car with machine guns. Held them at gunpoint. But it wasn't her guts I admired most. It was her heart. Her wisdom. The life she lived . . . She survived KKK attacks as a child in the South. She endured the sight of her father's accidental death. She conquered scoliosis to become a runner. Diana taught me a helluva lot more about life than I ever taught her about police work."

Mike usually tried to avoid large gatherings of police officers, but nothing was going to keep him from the funeral. He did, however, manage to get a seat without really talking to anyone else. Then, during the service, he found himself feeling raw and vulnerable. The feelings caught him off guard; he wasn't usually the emotional type. He had to work to keep from choking up. Once the service ended, he was hoping to leave as unobtrusively as he'd arrived, but he bumped into Williams on his way out.

"He came up and asked me how I was doing." Mike, on automatic pilot, summoned what had become his stock answer. "What I always say. 'Fine, thank you.'" They were making only small talk, but Mike felt weird. "I didn't feel like I normally felt." He couldn't put his finger on the feeling—a little bit of anger, maybe, but more like his mind could not stay focused on Williams's chatter and was instead trying to land on a thought just beyond his mental reach. It was most like the experience everyone has had at one time or another—when you see someone you know but for an instant you can't recall the person's name. There's a gap, a space in time, before the click of recognition. That's how it felt to Mike, although he was grasping not for a name but for a memory.

Mike took the feeling home with him. He had little to say to Kimberly or the boys. He was brooding, trying to make a connection to the nagging but unconscious thought. "I was up most of the night. I really wasn't sleeping." Then, without warning, it came. It gushed out in a rush of sounds and flashing lights. "I heard sirens blaring and people yelling." He was reliving the beating. He remembered standing at the fence and getting hit from behind. He remembered being on the ground, huddled, the blows coming down on him. "I remembered it in more detail than I ever remembered before."

New details surfaced. The most telling was this: Mike, as he was balled up on the ground, heard a voice in the cloak of darkness all around him. "Stop, stop, he's a cop!" It was a voice Mike recognized, a voice he knew well: Dave Williams's.

Kimberly awoke and found her husband shaken and agitated. Mike could not turn off the voice ringing around inside his head: "Stop, stop, he's a cop!" He told Kimberly what had happened, about this new information. And the next day he drove to the courthouse. He walked into Bob Peabody's office and told the prosecutor. Mike didn't exactly know what it all meant. Maybe Williams was a beater who called off the attack after recognizing him under the bulky hoodie and coat he wore. Maybe Williams was not a beater and had stopped the others from hitting him. Either way, Mike was convinced Williams was a key witness. He had a new twist to Williams's own jangle of *You know I know you, Mike.* It was now Mike's turn: *Dave, I know you know something.*

CHAPTER 13

Cox v. Boston Police Department

Bob Peabody faced Dave Williams. "Officer, at any time while you were there did you yell out, 'Stop. He's a cop!'?"

"Negative."

"Is that a yes or no?"

"No."

"You did not yell that out?"

"No, sir."

"You did not yell it out once or twice?"

"I didn't yell that out at all, sir."

"And you're sure of that?"

"Positive."

"And that is the truth?"

"Absolutely."

Peabody asked Williams the question five times. Minutes later, he tried a sixth. "You categorically deny that you uttered the words, 'Stop. Stop, he's a cop!'?"

"I never said that, sir."

The heated exchange came more than an hour into Williams's appearance before Peabody's grand jury. It was Friday, the first of December, and seven months into Peabody's investigation. Just as other witnesses had, Williams first checked in with a court officer stationed in an anteroom to guard the door. Williams then waited for Peabody to come out and get him. He walked into a room that was more an amphitheater than a courtroom, with three ascending rows of chairs where the twenty-three jurors sat. The bank of windows in the rear provided plenty of light, especially in the afternoon when the sun set. Williams took a seat in the flat area. His chair faced Peabody seated at a desk.

The prosecutor was getting nowhere. While he was uncertain whether Williams had actually delivered a blow or two, he was convinced Williams saw it all—"a bird's eye view," was how he put it.

Williams pushed back hard. "I have no idea what happened to Mike," he told the grand jury. "I didn't see anything of that nature, anyone striking Mike." Under oath, he elaborated on the skeletal accounts he'd initially given in his written reports. He said he'd bolted from his cruiser and chased the suspect named Jimmy "Marquis" Evans who ran from the Lexus's front passenger seat. "He was running. I drew my gun, and I told him, 'Get down. Get down. Get down,' and he did. And I ran up to him. I just kept him at gunpoint." He said he was in no position to see anything at the fence.

The account of a foot chase didn't square with the tight geography at the end of the cul-de-sac, but Williams was unflappable as he addressed any inconsistencies. When Peabody pointed out Williams had written in one January 30 report, "both suspects fell down a steep hill," Williams acknowledged the inaccuracy as a harmless mistake. "I put down 'steep hill' but I knew it wasn't—I couldn't tell whether the hill was steep or not."

Williams parried every incriminating remark others had made about him. He acknowledged saying at the hospital that it looked as if cops had beaten Mike, but insisted he was simply thinking out loud, putting words to what everyone in the room was thinking. "We were talking to Mike," he said. "We realized that something had happened that shouldn't have, basically."

Peabody brought up Craig Jones, who had said that Williams fingered his partner, Jimmy Burgio. Peabody had decided Craig's information had the ring of truth. First, he did not think Craig would ever concoct an incriminating statement against another officer. Moreover, the remark was not the sort of thing Craig or anyone would forget. "I'm asking you," Peabody said. "Did you ever tell Craig that you thought that your partner hit his partner by accident?"

Williams put his foot down. "No, I did not."

"Are you sure of that?"

"I'm positive of that."

"Absolutely sure?"

"I'm sure."

Williams explained Craig had misheard him. Sure, he'd brought up Burgio's name with Craig, but only to refute rumors already circulating in the immediate aftermath of the beating. He'd told Craig that Burgio could not have hit Mike because "he was with me." Burgio, he said, had run after Ronald "Boogie-Down" Tinsley, the other suspect who'd fled from the right side of the Lexus. Williams said he'd seen Burgio get out of the car. "When I turned around, when I had my suspect, he was there." Williams said Craig had misunderstood and twisted his words all around. He was actually vouching for Burgio.

Finally, Williams exhibited an impressive *inability* to identify any of the other officers at the dead end, despite Peabody's repeated efforts. He was comfortable discussing cops whose presence was widely acknowledged—Craig Jones, Joe Teahan, and Ian Daley, for example—but was stumped when it came naming any others.

"There were other cops there," he acknowledged.

"Did you know any of them?" Peabody asked.

"I can't recall exactly who was there."

Williams certainly appreciated the benefits of cops sticking together. In October he'd gotten word in one of the two excessive force complaints pending against him—he'd been exonerated. In the Internal Affairs inquiry, every other officer had backed his position that he'd struck the Dorchester woman only after she'd assaulted him. The eleven officers interviewed by IA either said they did not see Williams hit the woman or said that Williams restrained the complainant only after she had hit him. "Officer Williams was met with physical resistance while making a lawful arrest and used the minimum amount of force necessary to subdue Miss June Ivey," ruled the investigator for Internal Affairs.

By the time he was finished questioning Dave Williams, Peabody was deflated.

"Is there any information, Officer Williams, that you can give this grand jury that would assist them in determining what had happened to Michael Cox that night?"

"Just what I told you," replied Williams.

It was a command performance by an officer who, in the hours after the beating, had likely said too much and was now explaining it all away to return to the police fold.

While Bob Peabody was pursuing his grand jury probe into the Cox beating during the fall, New York's new police commissioner—the brash and high-profile Bill Bratton—was holding forth at Harvard Law School. Bratton, the ex–Boston police commissioner who'd once gone on a ride-along with Mike and Craig, had agreed to be the keynote speaker at a forum titled "Police, Lawyers, and the Truth: A Symposium." Bratton was called on to address the problem of cops lying to make an arrest or while testifying at criminal trials.

The police perjury, nicknamed "testilying," was believed to be a by-product of the stand-together police culture that was responsible as well for the blue wall of silence. They were, in effect, branches of the same tree. With testilying, cops lied usually to protect a case and ensure a conviction. With the police code of silence, cops lied to protect another cop suspected of wrongdoing. In both, it was all about us versus them.

"You cannot break the law to enforce it," Bratton began.

Testilying had moved front and center into the national dialogue, a hot public issue for much of the year due to the O. J. Simpson murder trial. Allegations of misconduct and perjury by the Los Angeles police were a centerpiece of Simpson's defense. With more than 150 million viewers watching the televised verdict, the former football star was acquitted on October 3 after his jury deliberated less than three hours.

In particular, Harvard Law professor Alan Dershowitz, one of Simpson's lawyers, had been outspoken about testilying, and, indeed, at the Harvard forum he reiterated his combustible claims that cops not only lie routinely but actually teach one another how to do it. Many police and police unions were livid with Dershowitz, and some police chiefs boycotted the Harvard forum because of him.

Bratton, however, had not. Dershowitz and other criminal law experts, he told the audience, "have said police perjury is pervasive. If you asked the police unions, they would say it is minimal. I think the truth probably lies somewhere in the middle.

"This is enough of a problem that we need to address it. We can't address it by ignoring it, and we can't address it by boycotting conferences like this one." The practice, he insisted, was basically well-intentioned. "Testilying is different from any other form of police corruption because it is usually unrelated to any opportunity for personal gain. Cops who testilie do so in the belief that they are helping to enforce the law . . . As the cops who testilie see it, they don't lie to convict innocent people, but to convict the guilty."

Bratton had chosen not to belabor the wholesale corruption behind the lying. But, truth be told, the made-up testimony—purportedly to give justice an edge—could indeed provide cover for the corrupt cop who was bad to the bone. Boston, at that very moment, had two veteran detectives who for more than a decade treated Roxbury and Dorchester as their own money store. Publicly, Walter "Mitty" Robinson and Kenny Acerra were known as street-savvy crime fighters often called on to help solve some of the biggest cases. They had press clippings saying so. The reality was that on the street they were a two-man crime spree. Lying routinely to obtain phony search warrants, lying routinely in court to cover up their actions, they shook down drug dealers for their money, drugs, and guns. Little did Bratton know, but as he spoke at Harvard, the two

Boston detectives, having grown sloppy, were about to be caught. Federal investigators documented fifty-six cases where the two had shaken down suspects illegally. They were eventually charged with stealing more than $250,000 in cash, drugs, and guns.

When Bratton was finished, the Boston police union was not at all happy with him. The union's president angrily told the *Boston Globe* the police commissioner's views on testilying were "incredible." The union official, a twenty-seven-year veteran, even denied testilying existed. "I went to court an awful lot and I can never remember any problems of this kind."

The denial had a hollow ring to it—just like denials of a blue wall of silence.

Bob Peabody ran into a wall with Dave Williams. The unproductive standoff was emblematic of the overall lack of progress in his investigation. In seven months he'd put five Boston police officers and another two municipal police officers before the grand jury for questioning. With Farrahar, he'd interviewed twenty-two police officers and munies in the offices of the Anti-Corruption Unit in Fort Point Channel. He'd worked up a theory of culpability revolving around the big three—Burgio, Williams, and Daley—with Burgio as the principal assailant. In his reconstruction, Burgio exited the passenger side of Williams's cruiser to find Cox standing right in front of him at the fence. "My theory was he was the first at the fence and brought Cox down." Peabody found unofficial confirmation in Jimmy Burgio's choice of defense counsel. Burgio showed up for his interview in July accompanied by Thomas Drechsler, one of the smartest attorneys around, who often represented police officers in trouble. "To me, that confirmed Burgio's the guy," Peabody said. "Why else would he have the top attorney?" It was, of course, rank speculation mixed with gallows humor, but Peabody was only half joking.

In truth, Peabody had not made any real headway toward charging anyone. Burgio took the Fifth and refused to answer any questions in July. When Ian Daley showed up for his interview one overcast evening in August, his tape-recorded session lasted all of three minutes. Just long enough for Daley to give his name, rank, police ID number, and refusal to cooperate. "I respectfully decline to answer on the advice of counsel," he said, citing his constitutional privileges against self-incrimination.

Then Joe Teahan and Gary Ryan invoked their rights against self-incrimination—a move that confused and surprised Peabody. He'd not suspected either in the beating—indeed, they'd attended to Mike—but once they joined Burgio and Daley as the only officers taking the Fifth, he was tantalized and began thinking they had information. "They might have heard something. I wanted to get that." He sought immunity for both—a court order protecting them from prosecution in exchange for their testimony. But once under oath, Teahan and Ryan had nothing helpful to say; they said they'd arrived after the beating. Peabody was flummoxed. "I thought, 'What? We went through all this for nothing?'" He realized he should not have immunized Teahan and Ryan without a proffer—a preview of what they had. "That was just also a learning curve for me."

For all the effort, his investigation had not only failed to advance the case but it actually turned up less than Internal Affairs. Richie Walker, for one, was in full retreat. He'd told Jim Hussey he'd seen another officer running after Mike toward the fence. When Walker met with Farrahar, he described briefly seeing Mike run toward the fence, but made no mention of seeing anyone behind Mike.

With Jim Hussey, Walker had said Mike's beating was the talk of the Roxbury station house later that night. With Farrahar, Walker flat-out denied anyone later discussed the beating. Walker was asked: "Did you have any conversation with either Officer Daley or Officer Jones or Sergeant Thomas in regards to the injuries that Michael Cox sustained?" "No, sir," Walker replied. "I just asked how is he doing." Peabody, unable to study any of the IA records, had no way of knowing about Walker's about-face. Walker was like so many of the other witnesses he'd questioned; he saw nothing.

The grand jury also never heard from Bobby Dwan. Dwan

told Hussey he'd seen two uniformed Boston cops, one white and one black, in a commotion by the fence. He said when he arrived he'd seen both Teahan and Ryan moving about the dead end. But Peabody and his investigators never interviewed Bobby or Kenny Conley. Both would have cooperated had they been summoned, but the summons never came. For his part, Kenny had put the contentious end to his last session with Hussey behind him. He wasn't the sort who brooded. He'd resumed his life's routines—his night shifts in the South End, his basketball playing. Kenny certainly heard the same rumors everyone was hearing—about Burgio and the others— but he wasn't paying much attention. It didn't involve him. "I pretty much forgot about it."

Somehow, Kenny and Bobby had fallen through the investigation's crack; Peabody did not realize Kenny and Bobby were at the dead end. He was not allowed access to their interviews with Internal Affairs, and he also apparently never received their written reports. "We somehow didn't get our hands on that," Peabody said later. The presence of Kenny and Bobby at Woodruff Way, Peabody said, "never came up."

It all meant Peabody had less to work with, and by the time of his face-off with Williams on Friday, December 1, he'd lost the zip and confidence he'd started with. Sunday evening from his home, he typed an e-mail to Ralph Martin. The session with Williams, he told his boss, was "not what I hoped for."

Peabody continued, "He stood his ground when confronted with damaging statements he supposedly made to others who have testified. He flatly denied yelling, 'Stop, he's a cop!' when Cox was getting hit and denied telling Craig Jones later that night, 'I think my partner hit your partner.'"

Williams, wrote Peabody, "said he saw and did nothing other than chase the murder suspects."

If it all sounded defeatist, Peabody wasn't ready to fold yet. Williams, he said, was the key. "I think he saw it and has convinced himself that this is the story he is going to give, or he really didn't see what happened, believes that his partner was probably involved and has decided to protect him as best as he can." Peabody said he had an idea. "It is time to confront Williams. Lay our cards and theories on the table and see what he says. There are sufficient contradictions now on the record to smoke him out if he's hiding it." Peabody wanted to arrange a meeting with Williams and his attorney.

He wanted to let it all hang out. It would be a Hail Mary.

Mike jerked upright and leaned over the back of the couch. Groggy with sleep, he took a split second to get his bearings. Then he carefully pulled back the curtain to peek outside. He was convinced he'd heard something. But Supple Road was quiet. He looked up and down his street. He saw nothing, at least not what he was looking for. His unmarked police cruiser sat in front of his house, untouched.

Mike had taken to sleeping on the living room couch after finding the first tire slashed one morning when he left the house for work. "I was trying to catch them." But he hadn't, and over the next few weeks, the other three tires were cut up. His car was clearly targeted; it was the only one on the street that was hit. Mike was certain cops were the culprits, cops who'd adopted yet another technique to communicate what they thought of him, "that I was becoming some type, you know, of rat." In the police world, tire slashing was known to be one way cops expressed displeasure with one another.

The harassment started as officers received subpoenas in late summer to appear before Peabody's investigative grand jury. Mike's return to work was not going well. "I'd just walk into a room and, you know, people look at you like you're dirt." Mike listened to some commanders reassure him his beating was unacceptable, but the talk was empty, particularly when he could just look around and see actual suspects still on the job. No one had yet been disciplined in any way, despite all the lies the investigation had already established. Some were even promoted. Sergeant Dan Dovidio, for one, rose to the rank of sergeant detective. Not only that, he was transferred to Internal Affairs. It couldn't have gotten any more bizarre—the supervisor who'd retreated to the police station when nearly every cop on duty was racing to the shooting at Walaikum's, the supervisor who'd told Williams and Burgio at Woodruff Way to lie about riding in the same cruiser, was now seen by the commissioner as having the right stuff to uphold the department's integrity and standards of conduct.

It was all a bit hard for Mike and Kimberly to take. "Life for me became more and more difficult," he said, "and I just didn't understand, you know, why? What did I do to create all this hostility?" Mike had several times changed their telephone number and had it unlisted, but that didn't matter. The crank calls continued, albeit with periodic breaks. Then one night a crew of Boston firefighters and fire trucks arrived in the middle of the night, apparently summoned by a false report that the Cox house was on fire.

Now there were the tires. When Mike lay back down on the couch, a video camera, pointed out of the living room window, continued making its slow, whirring sound. The camera was aimed straight at Mike's cruiser. The car's shadowy image was displayed on a monitor attached by cables to the camera.

Farrahar's Anti-Corruption investigators had installed the camera. It was a primitive setup, requiring Mike to actually "do a lot of rewinding and setting up of this equipment, turning it on and off." Kimberly was unimpressed; the setup, she said, was a "joke. It's like it was something from 1950s. The picture was so unclear, it was just basically fuzz." It seemed so ineffective. "See a picture? I mean, looking at it, I couldn't make out much of anything; maybe shadows." Within a couple of weeks, she and Mike had had enough of the Boston Police Department's putative high-tech capabilities. The camera was more a nuisance than anything else, and they insisted it be removed. Mike would keep trying to capture the slashers on his own.

Kimberly was put off by the whole thing. To her, the clunky surveillance equipment was a token, even patronizing response. "I didn't think that was a serious attempt for them to find out who was doing this." In fact, it became a symbol for how the couple now viewed the overall investigation—halfhearted, bungled, and wanting.

By early fall, Mike had seen enough. He'd always believed in the system, but he now reached the conclusion the system had fallen short. "I was failed by the police department." Just as he was on his own when it came to the tire slashers, Mike decided he was on his own in the search for justice. "I had to do something," he said, "regardless of what the DA's office or the police department was going to do." The continual harassment, rather than a deterrent, had become a prod. "I decided, along with my family, that I needed to find out, you know, who was involved, who did this." Mike realized he was going to have to take matters into his own hands.

He hired Steve Roach and began meeting regularly with the attorney. Then, in late fall, as Bob Peabody was unsuccessfully pushing Dave Williams to come clean, and six days after Bill Bratton's appearance at Harvard Law School, Mike sued. He sued his fellow cops, his police department, and his city. He said his civil rights were violated when Boston police officers repeatedly beat and kicked him until he blacked out. He charged that David C. Williams and Ian A. Daley witnessed the attack, did nothing to stop it, and then left him injured and unattended on the street. He took on the police culture of silence and said the two officers joined others in a cover-up and failed to report the assailants. And Mike took on the Boston Police Department. The department, he said, "fails to investigate allegations of misconduct by police officers, fails to properly supervise police officers and fails to properly train police officers and their supervisors after having prior knowledge of multiple incidents of misconduct, especially against young black male suspects, other powerless citizens and plainclothes officers."

Mike was in metamorphosis—moving from cooperating victim in others' investigations to aggressor in the quest to hold his assailants accountable. He'd been a punching bag that night at the fence, and he'd felt like one ever since. It was now about "my self-esteem" and "my family." He could no longer be a bystander.

"It was humiliating what happened to me," he said. "There's no reason to treat anyone like that. And then to just leave them. And if they do it to me—another police officer—would they do it to another person if they got away with it?

"What's to stop them? Who's to stop them?"

On December 31, an estimated one million revelers turned out for Boston's First Night activities. It was the twentieth year the city hosted a long day's celebration into the night, featuring towering ice sculptures, a parade, puppet shows, music concerts, and, at midnight, a fireworks display over Boston Harbor. For the occasion, more than two hundred Boston police officers and ninety-one police cadets were deployed to keep the city's record of a festive and peaceful New Year's Eve intact. "We're going to keep this a safe and enjoyable way for people to celebrate," Mayor Menino promised beforehand.

Mike Cox was not feeling particularly celebratory or safe. He'd filed a federal civil rights lawsuit against his police colleagues and his department. He knew he was stepping way out of line. "I have accused them of things which I don't necessarily know anybody else in the police department has ever done before." The claims could cost them their jobs and monetary damages, and "it could send them to jail."

Taking legal action may have brought some satisfaction, but Mike now wrestled with the fear factor. For Mike, it went like this: He'd become a troublemaker, and the quickest way for those troubles to end was "by me not being on this earth or being killed." That was the way Mike Cox's year ended—believing his life was at risk. It wasn't the unfounded fear of an outsider. Mike was one of them. He'd been a cop for six years and knew the score. He understood completely that his lawsuit meant that he was locked in combat against the police culture, and, by taking it on, he had become the enemy.

PART III

Justice Denied, Then the Trial

CHAPTER 14

The White Guy at the Fence

By the time he appeared before Bob Peabody's grand jury in December, Dave Williams was likely feeling more emboldened than ever. He'd come up with explanations for the incriminating utterances he'd made the night of the beating, had stuck to them during repeated questioning, and had not faced any heat. In fact, the worst that had happened was being named in Mike Cox's civil rights lawsuit, and Williams let Mike know what he thought about that.

He called Mike after receiving notice of the lawsuit. "I got something in the mail."

"Yeah, I assumed you would," Mike said.

Mike was at his desk in the Internal Affairs office. Williams said he wanted to see him right away. "Well, I'm here," Mike said.

Without hesitation, Williams marched into Internal Affairs. He asked what Mike was up to suing him and the department. He reiterated his talking points from their chance encounter in Franklin Park: You know I know you, he said.

Yeah, Mike said, he knew all that.

"I hope you don't think I hit you," Williams said.

"Just do the right thing," Mike said.

Williams grew agitated. Right thing? Right thing? "Fuck everybody," he said.

"You should fuck everybody," Mike said. He told Williams he did not owe anyone anything. "So why cover for them?"

They talked in circles and then Williams left. It took Mike a few minutes to fathom the strangeness of the moment: a prime suspect in the beating challenging the beating victim in the offices of Internal Affairs. And it turned out Mike's boss, Deputy Superintendent Ann Marie Doherty, had even spotted Williams. But she did not intervene and merely told Mike later that William's presence was not appropriate.

Williams had dug in and was now on a roll. It seemed even when he lost, he won. He got the result of the last excessive force case pending against him, the one involving teenager Valdir Fernandes, who claimed to Internal Affairs and in a lawsuit that Williams slapped him around on his porch. Williams learned he'd been found guilty of physically abusing the teenager. Two officers had backed Williams, brazenly submitting reports that were virtually identical, except for signatures and a few token changes in wording and grammar. But there was a glitch; the identical accounts didn't quite line up with Williams's. The two noted that Williams, after the boy spit near him, "grabbed Valdir above the jacket." Williams, in his report, never mentioned touching the boy at all. The discrepancy was enough for the Internal Affairs investigator to rule that Dave Williams's account was "less credible" than the boy's. Making matters worse were the photographs of the boy showing "bruises consistent with being slapped in the face."

For once it looked bad. But appearances were deceiving. Once again, even an adverse result proved much ado about nothing. No disciplinary action was taken against Williams, and, eventually, the city paid Valdir Fernandes $7,000 in a settlement to end the lawsuit against the police.

Two months after making the Hail Mary proposal to his boss, Bob Peabody got his wish in early 1996 to take a final run at Dave Williams. He and Lieutenant Detective Paul Farrahar arrived

at Williams's lawyer's office, located in a building near Quincy
Market, the popular shopping area in the shadow of the elevated
Southeast Expressway that cut through Boston's downtown, on a
gray February day.

Waiting in a windowless conference room was Williams and
his attorney, Carol S. Ball. Peabody had known Ball for years; she
was a former assistant district attorney and he'd long admired her
feisty, high-energy advocacy. "I could talk to her and vice versa,
and so I asked for this audience." The meeting had come together
quickly. Ball had been named to the bench and would soon leave
her practice to become a Superior Court judge. She would have to
quit representing Williams—and, for that matter, all her clients.

The ground rules were simple: The session was unofficial and
off the record. Peabody's hope was to persuade Williams to come
around. He and Farrahar decided beforehand to do the "good cop,
bad cop thing," with Peabody playing the role of bad cop. His ap-
proach would be: "C'mon, Dave, here's your chance. Tell us what
happened. I think you know more. I think you saw more."

The prosecutor began by explaining why he'd asked for the
meeting. He told Williams his story about chasing Marquis Evans
was contradicted by two other police officers. He urged Williams
to "come clean rather than protect his partner that night."

But Williams was unmoved. "He was terse," Peabody said later.
"He's a big man. I'm a big man, too, but I'm not as big as he is, not
as wide. He just said, 'That's all I know.'" It was as if Williams was
saying: You're not intimidating me with your hardball stuff.

Peabody quickly saw the last-ditch tactic was a bust. "We weren't
getting anywhere," he said. "I just couldn't crack him." Reluctantly,
he threw in the towel.

"It was a short meeting."

Peabody's last hope was Ian Daley, who initially struggled and
asked Jimmy Rattigan for advice about what to do, but then clammed
up. Like Jimmy Burgio, Daley invoked the Fifth Amendment when
summoned to meet with Anti-Corruption investigators during the

summertime. But by early March 1996, Peabody was talking with
Daley's lawyer, trying to turn Daley into a cooperating witness. Pea-
body believed Daley had had a "ringside seat" at the dead end and, if
Peabody could flip him, Daley would make an ideal witness.

Peabody and Daley's attorney, a former local prosecutor named
Tom Hoopes, went back and forth. Hoopes was a seasoned practitioner. Before making any deal, he wanted Peabody to reveal the
evidence gathered against his client. He demanded that Peabody
actually let him read the transcripts of the secret grand jury testimony.

Hoopes wanted, in short, to see Peabody's hand in what
amounted to a legal poker game. He was doing what most any
defense attorney would do—consider cutting a deal that would
minimize or even eliminate criminal liability in return for the client's testimony. His primary legal responsibility was protecting
his client from criminal charges. But in this instance there was a
second and nearly equal concern weighing heavily against any form
of cooperation. Cops don't want to testify against other cops; to
do so was tantamount to professional suicide. But there was a scenario that could justify Daley becoming a witness: if other cops had
already started talking and were hurling accusations against him.
Under those circumstances, Daley could explain he hadn't run to
the courthouse and been the first cop to make a deal; he'd done
so reluctantly and only to defend himself. Most cops would understand that.

Jockeying with Peabody, Hoopes had to keep all this in mind.
He needed to know what Peabody had on his client, and, to get
that, he indicated that Daley was considering a deal. What, if anything, had other cops said against his client? Had anyone gone so
far as to finger Daley in the beating? Hoopes needed answers. He
swung by Peabody's office on the first Monday in March, and the
two lawyers discussed the Cox investigation.

Peabody, afterward, wrote his supervisor an e-mail. "He has revised his latest demand to view GJ testimony re: his client's (Daley)
involvement," he wrote. "Instead, he asks that I give him an oral

summary of what evidence (to date) impacts his client and/or how he is involved." It may have been wishful thinking, but Peabody said from talking to Hoopes, "I got the impression that Daley may want to cooperate."

If true, Peabody saw a huge upside. Hoopes, in the give-and-take, said his client would be able to get Peabody "halfway there" in solving the assault, but that he was unable to "name names." Peabody wasn't sure what this meant. Did it mean Daley was not able to provide a specific blow-by-blow—saying which cop hit Mike first, dragged him from the fence, and so forth—but that Daley had seen the beating and could identify the assailants, if not by name then by description? Or did it mean Daley had not witnessed the actual beating, but, seconds afterward, saw the cops who'd done it standing right there talking about their brutal handiwork? Peabody may not have known the answers, but he was sure of one thing: Daley was a valuable witness whose cooperation he wanted.

But Peabody was also aware of the risk of telling all. "The downside," he wrote, "is that he will see our cards and, now knowing he's safe, keeps silent." The prosecutor was at a crossroads, desperate even. "I think we should comply with his request," he recommended, and got the okay.

Peabody disclosed to Hoopes that no one was accusing Daley of beating Mike and he had no evidence to charge him. Weeks went by, and Peabody heard nothing back. In early April, after nearly a month, he called Hoopes. Was Daley going to play? He pressed for an answer. Then Hoopes delivered the bad news—thanks, but no thanks. Peabody reported the devastating kiss-off in another e-mail to his boss, the DA Ralph Martin. "Daley said NO despite having the unusual benefit of knowing almost everything our investigation has learned to date about the events of January 25, 1995." Peabody's worst-case scenario had indeed come to pass. He'd been outmaneuvered—baited, in effect—by the long shot he'd get Daley to cooperate. For Daley, it was a no-brainer. Why stick his neck out? In a game where justice had taken a backseat, he had no incentive to break his silence. Peabody had nothing on him.

"The test of loyalty oftentimes on police departments," a Boston police official once said, "is, number one: Will you lie for me? And if you won't lie for me, will you at least be silent?" In short order, the frustrated Peabody had suffered a wrenching one-two punch from Dave Williams and Ian Daley, the two cops he considered key to his prosecution. "You could tell Williams and Daley knew more," he said. "I couldn't get either of them to take it to the next level." Peabody was not sure what more he could do.

"We've sort of run out of gas."

While Bob Peabody was realizing a year into the investigation that he was spinning his wheels, his colleagues happened to be on the losing end of a courtroom battle against another Boston police officer accused of corruption. The cop had been charged with stealing $6,352 in cash from a wallet a civilian turned in to police after finding it on the lobby floor of her apartment building. The officer insisted he had not stolen the money. He'd left the wallet on the counter, he said, and while he was attending to another matter, an unidentified man came into the station and took the cash. The officer certainly seemed caught in a bind prior to the trial—trouble of his own making. Investigators uncovered two incident reports he'd written. The first recorded receipt of the wallet, the $6,352, a Hermès watch, and an airline boarding pass. The second, replacing the first, mentioned a wallet, the watch, and the boarding pass—but no cash. Taking the stand to defend himself, the officer testified he wrote the second report not because he stole the money but to cover up the embarrassment of leaving the wallet on the counter. He was guilty of a lie, not larceny. The jury agreed, and, on a Friday in late April, the officer was acquitted.

The media called the verdict a "stinging defeat to Suffolk County prosecutors," a reference to Ralph Martin's losing track record against Boston cops suspected of wrongdoing. Earlier that year, a judge had thrown out the case charging a cop with raping a prostitute with a nasty public rebuke, and now this losing trial verdict. Not surprisingly, Tom Drechsler represented the cops in both cases, the lawyer of choice for cops in trouble. Jimmy Burgio's attorney was fast establishing himself as Martin's nemesis.

In the weeks following this latest setback, Martin entered into discussions with the U.S. attorney's office in Boston about federal prosecutors taking over the Cox investigation. Peabody had not been able to uncover the kind of evidence he believed necessary to convict cops responsible for Mike's injuries. Bob Peabody needed eyewitnesses. "We didn't have anybody saying, 'It was him, it was him, it was him.'" It was no longer enough to build a circumstantial case, not in the hothouse atmosphere of cop cases. "You want slam-dunk evidence," Peabody said. "We certainly didn't have it."

Having come up empty, Peabody began directing his focus elsewhere. He was assigned to prosecute an electrician charged with involuntary manslaughter after three children died in a house fire sparked by the electrician's faulty wiring. It was a complicated case, and Peabody started preparing for the trial in early January, when he would argue that the electrician's work in a basement apartment was so atrocious and careless he should be held criminally responsible. Peabody's career also took an important turn. Working alongside federal prosecutors in the Charlestown "code of silence" racketeering case, he'd made a strong impression. He was offered a job as an assistant U.S. attorney and would be leaving Martin's office.

Nonetheless, Bob Peabody continued to chase a lead or two in the Cox case. One involved the municipal police officers. He got word one munie was overheard telling another, "You shouldn't have hit him so hard." He conducted interviews, but decided quickly the line was locker room banter. "Gallows humor," he said. The remainder of 1996 was mainly about preparing for the electrician's trial— his last as an assistant district attorney—and boxing up his Cox files for federal prosecutors. Maybe the feds would have better luck breaking down the police culture and getting cops to talk.

To help out the U.S. attorney's office, Peabody typed up a four-page, single-spaced memo, "Status of Cox Investigation." It was a confidential summary of the failed grand jury investigation. "This investigation is stalled," he wrote. He covered Mike's beating and then described what he'd done to try to solve it, including his confrontation with Dave Williams and the unsuccessful negotiations with Tom Hoopes about Ian Daley. "We have questioned (sometimes twice) every person present that night; persons who would have any information about the Cox beating." Peabody didn't realize it at the time, of course, but he was wrong about having gotten to everyone. He had missed that cops from the South End station were also there, namely Kenny Conley and Bobby Dwan.

The Cox case would soon become the worry of federal prosecutors, but Peabody seemed reluctant to let go completely. He recommended the DA's case be kept in an "active" status just in case one of the police officers testifying before the grand jury would "change his mind whether because of what's happening internally at BPD or for reasons of conscience."

The notion of a cop doing an about-face as a matter of conscience seemed fantastical, given his experience with the police department. Peabody then had a thought that seemed to hold out greater promise. It involved the four men in the Lexus: Smut Brown, brothers Tiny and Marquis Evans, and Boogie-Down Tinsley. Throughout his Cox investigation, Peabody had purposefully stayed away from them, not wanting to risk in any way the integrity of a murder case. But the four were all headed for trial in late October for the murder of Lyle Jackson. The trial, observed Peabody, depending on its outcome, "may shake one or two of the lesser culpable defendants into our camp."

Wouldn't that be a twist—cooperation not from the cops but from the criminals?

"Hearsay," he continued, "says that the rear left passenger saw the whole thing and can ID more than one white cop beating Cox."

Peabody was talking about Smut Brown.

The four were being prosecuted by the DA's office on the theory the murder was the result of a joint venture, a legal term describing a situation in which one person is held responsible for another's crime if evidence proves they worked in concert and shared the same mental state. Smut Brown, even though he did not pull the trigger, could be proven guilty of murder if the prosecutor could show that Smut and the other three, as a team, killed Lyle.

Smut was represented by a court-appointed attorney named Robert L. Sheketoff. Sheketoff, forty-seven, had come of age in the late sixties, attending an almost all-black high school in Hartford, Connecticut, and then Brandeis University near Boston, where he majored in politics. For the two decades following his graduation from Yale Law School in 1975, the curly-haired, bespectacled, and bookish-looking trial practitioner had represented all varieties of the accused, from small-time drug dealers like Smut to the underboss of the Boston Mafia, Gennaro J. Angiulo. He'd done so without fanfare; while many lawyers regularly sought media attention, Sheketoff couldn't be bothered. Despite his low profile, he was the legal profession's inside secret, a brilliant workhorse determined to hold the government to the test during a criminal trial, admired by other defense attorneys, prosecutors, and judges alike for his laser-like skill to instantly assess a witness's testimony for holes and inconsistencies.

"Trials are like plays—ad-lib plays," Sheketoff once said, "and you absolutely have to listen to every word. The most important thing you can do is listen to the witness."

It didn't always happen with clients, but Sheketoff had grown fond of Robert "Smut" Brown. He didn't think Smut came close to being a cold-blooded street criminal capable of killing. He'd seen plenty of those types in his career. He saw instead a charmer, with a spark in his eye, who grew up in tough surroundings and had made plenty of bad and stupid choices. He loved his kids. Drug dealer? Yes. Killer? No.

"He had a very good side to him," Sheketoff said.

With Smut, Sheketoff's goal was to show the jury Smut had nothing to do with the fatal shooting; that Smut was as surprised by the outburst of gunfire as the patrons fleeing Walaikum's; that the killing was sudden and not premeditated; and, finally, that Smut was outside when all hell broke loose. In sum, Smut did not share the mental state required for a joint venture and was not guilty in the tragic slaying of Lyle Jackson.

While preparing for trial, Sheketoff sometimes found it hard to focus solely on the murder. The reason was the Cox beating. It may have been ancillary, but the beating was like a giant elephant in the room. He thought he'd seen it all. "Maybe this is too cynical, but I don't think there's anything particularly unusual about police beating up suspects, you know, going overboard." But Cox was different. "What's hard to comprehend," he said, "is when it turns out it was a police officer who was almost killed, but the police solidarity is such that no matter how old, how young, male, female, black, white, the police officers—they all keep quiet about what happened. That's scary."

Smut added to Sheketoff's curiosity about the beating. From the time of his arrest, Smut told his public defender and then Sheketoff about witnessing the cop pile-on. "Long before the trial, I knew from Robert that he had seen the beating."

The murder trial began on October 28, an overcast, 60-degree day, with the jury taking a bus ride to see the crime scene at Walai-kum's and travel the more than ten miles of the chase through Roxbury into Mattapan, and ending at the dead end of Woodruff Way. The restaurant was so cramped jurors were taken inside in three groups of five, five, and six. Throughout, the prosecutor served as tour guide, pointing out relevant sites—the locations, for example, where the men tossed the guns out of the Lexus during the car chase. When they reached the dead end, he urged jurors to study the layout. Looking around, Sheketoff, for one, was brought up short. "Wow, this is pretty small," he said. He found himself wondering again about Mike Cox. "I couldn't imagine how a couple dozen cops were there, and no one saw what happened?" He didn't buy it.

The next day Lyle's mother, Mama Janet, took the witness stand—the first of twenty-six witnesses who testified during the eleven-day trial. It was a good call by the DA to start with her. She vividly brought back to life for the jury the murder victim, her twenty-two-year-old son, the father of a five-year-old-boy, working at a department store and trying his best to make his way in the world. She described how she got a telephone call in the middle of the night and ran around the corner to Walaikum's, where Lyle, riddled with bullets, was being loaded into an ambulance. Her son died six days later.

While the prosecutor was trying all four men as participating in a joint venture, he still had to settle on two as the actual shooters. From the witnesses police had assembled, he concluded the gunmen were Tiny and Marquis Evans. The proof came mostly from the one witness the DA's office determined was most reliable—a twenty-eight-year-old Jamaican named Alton Clarke. Clarke had been a constable in Jamaica prior to moving to Boston. Clarke had told police he saw two gunmen run from the restaurant and climb into the Lexus. He said they climbed into the car's front seats. The two shooters, therefore, were the driver and front-seat passenger—and that meant brothers Tiny and Marquis.

Smut was surely helped by Clarke's testimony; for starters, Clarke got it right and wasn't calling Smut one of the two shooters. But it got even better. Clarke went on to say that when the two gunmen jumped into the Lexus, he saw the other two men—namely Smut and Boogie-Down—were already in the backseats.

Sheketoff was delighted. In fact, Sheketoff learned prior to trial that the government had only one witness to argue Smut was part of a murderous joint venture, a friend of Lyle's who claimed Smut pointed out Lyle Jackson seconds before the shooting started.

Marcello Holliday began by telling the jury he and Lyle socialized at the Cortee's Lounge to "mess with women," where he drank "one or two beers." They headed to Walaikum's after closing time. He said he saw Smut huddled with the Evans brothers and then overheard Smut "whisper something to Tiny and Marquis." He quoted Smut saying, "that's one of them right there," indicating Lyle at the food counter. Following those words, Holliday said Marquis pulled out a gun.

Finally, when Holliday was asked to describe what Smut was

wearing, Holliday answered a tan Pele jacket—Pele being "the name of the jacket that was written on the back."

Sheketoff was quick to interrupt. "What was the answer to that?" Smut was not wearing a tan coat inscribed with the soccer star's name. He had been wearing a brown leather jacket the night Lyle was killed.

Holliday tried to take it back, saying he actually wasn't sure what Smut wore. But the damage was done. Sheketoff had emphasized the discrepancy for the jury.

Sheketoff sat listening intently as Holliday finished and, standing to cross-examine him, felt the testimony from the government's single witness against his client had turned out to be "manna from heaven." He challenged Holliday's math on his drinking. "One or two?" Was that credible for Hip-Hop Night at the Cortee's when he was on the prowl for girls? He asked Holliday more questions about the "Pele" jacket to further drive home the witness's foul-up. Most important, Sheketoff pounced on the word Holliday had chosen to describe Smut talking to the Evans brothers: whisper. Walaikum's was packed that night. Even if Smut had whispered something, how could Holliday have gotten it right with all the noise? In relatively quick fashion, Sheketoff showed the jury Marcello Holliday's testimony was not evidence it should rely on.

In presenting his case, the prosecutor called eleven Boston police officers to the witness stand—officers like Jimmy Rattigan, who described crashing his cruiser to avoid colliding head-on with the fast-moving Lexus, and officers like Roy Frederick, the off-duty cop who retrieved one of the weapons near his front yard. Critical in any murder prosecution was calling the arresting officers to the stand. The jury therefore heard Craig Jones testify about capturing Tiny Evans on the left side of Woodruff Way. Richie Walker took honors for Smut Brown, even though, in fact, Kenny Conley had caught Smut. But because of flawed police reports crediting him, Richie Walker was the cop who took the stand and promised to tell the truth. He told the jury how he ran between the Lexus and a police cruiser, stepping over two fallen suspects, Boogie-Down and Marquis, and then hustled through a hole in the fence to take up after the fleeing Smut Brown.

Jimmy Burgio was the officer who took credit for arresting Boogie-Down, but he was nowhere to be found. Burgio, due to the Cox investigation and the fact he'd taken the Fifth, was off-limits. The prosecutor therefore summoned Dave Williams to speak for his partner, Burgio, and about the arrest he'd taken credit for—Marquis Evans.

It was late in the day. Smut sat at the defense table next to Bob Sheketoff. Williams was the last of a string of officers called to testify. Dressed in uniform, Williams strode into the courtroom and made his way toward the witness stand. Smut watched the tall, broadly built cop with the mustache settling into his chair. His mind began spinning. Smut instantly recognized Williams from the fence. Williams was one of the cops he'd watched haul Mike Cox down and beat him. Smut couldn't hear Dave Williams's voice answering the prosecutor's questions—it was all white noise. His mind was in a paroxysm, crazy with the recognition. He leaned into Sheketoff.

"That's the guy," he said. "That's the guy."

But Sheketoff didn't hear his client. He was listening to Williams testify—and was in the midst of his own epiphany. "I'm thinking I know who beat up Mike Cox."

He found unbelievable Williams's account of him and Burgio apprehending two suspects in front of the Lexus. For one, it was contradicted by Richie Walker, who said he'd stepped over Boogie-Down and Marquis lying facedown between the Lexus and Williams's cruiser. Then Williams went on to say he'd chased Marquis fifteen to twenty yards and never knew about a fence with a hole in it encircling the dead end. The distance Williams described would have taken him through the fence and down the hill.

Sheketoff was stunned. It was a story where Williams, conveniently, ended up as far away from the Cox beating as possible. When it came his turn, Sheketoff rose and immediately went after the Boston police officer. He began by bringing out the fact Williams was caught lying by Internal Affairs about the cruiser he and Burgio rode in.

"It was just considered a minor infraction," Williams said.

He then went for the jugular and challenged Williams about Cox. Pointing to the area on a map of the dead end where Cox was beaten, he asked, "You and Burgio weren't over here?"

"No, sir," Williams said.

"Did you see anyone in that area over here behind your cruiser, in this area?"

"No, sir."

"Did you see any altercation that involved Officer Cox at any time?"

"No, sir."

The prosecutor began objecting, and the judge called a time-out. He had the lawyers walk to his bench for a conference outside the earshot of the jury. What, he asked, was Sheketoff up to? "I heard Officer Walker's testimony," Sheketoff said. "So it seems to me that *this* guy was not out chasing anybody. He was out doing something else."

But what's the relevance? The judge warned Sheketoff. "I'm not going to have this a trial of whether or not Mr. Cox was injured rightfully or wrongfully by police officers," he said. "This isn't going to be a trial of Mr. Cox's problems."

Sheketoff had indeed wandered off track. The case at hand was murder, not justice in the Cox beating. But he was outraged and unable to stop himself from momentarily substituting the public's interest over his client's. "Not that it really was any of my business," he conceded later, "but I'm also a citizen of the Commonwealth."

While Smut and the others were on trial for the murder of Lyle Jackson, the U.S. attorney's office in Boston began assuming control of the investigation of Mike's beating. Bob Peabody and Lieutenant Detective Paul Farrahar went over to federal courthouse, a worn, granite and limestone tower built during the Depression in downtown Boston, as part of the transition, to meet the federal prosecutor and FBI agent heading up the new probe. The Cox case had made it to the big leagues—the top of the investigatory food chain, the U.S. attorney, District of Massachusetts, U.S. Department of Justice.

The case was assigned to S. Theodore Merritt, a forty-four-year-old assistant U.S. attorney, originally from New York City. His given name was Stephen, but he went by Ted. Merritt had attended Harvard College and then Villanova Law School. He went to work for the federal government right away, in early 1978, after passing the bar exam. He began as a staff attorney in the Civil Rights Division in the Department of Justice. Seven years later, in 1985, he moved to Boston to become a prosecutor, first in the Criminal Division and then in the Public Corruption and Special Prosecutions Unit.

In court, the dark-haired Merritt often came off as humorless and all-business. He wore a neatly trimmed mustache and was average in height and build—standing no more than five feet, nine inches tall—but loomed larger, given his intensity. Some opposing attorneys found him menacing, a prosecutor who always made it clear he was acutely aware of the enormous power of the federal government and how it could upset people's lives. By the mid-1990s, specializing in public corruption, he'd made his mark as a prosecutor who mostly prevailed in court, and when he did, took no prisoners.

His cases included those of the ex-police chief of Winthrop, a town just north of Boston, who for more than a decade accepted $70,000 in bribes from the operator of video-poker machines; the Massachusetts state trooper who, while on duty, beat up a man outside a bar and lied afterward to cover it up; the guard in a county prison who threw a cup of boiling water on an inmate, in restraints, who was awaiting trial for the rape and murder of a young girl. In each, Merritt either won at trial or got the suspects to plead guilty.

During the meeting with Peabody, Ted Merritt and his investigators were given a quick rundown of the stymied investigation. The targets were identified—Jimmy Burgio, Dave Williams, and Ian Daley—as was the theory that Mike Cox was assaulted by at least one and possibly three cops. In the big leagues that belonged to the "feds," the main strategy to break a logjam was to find a cop to "squeeze." Best case scenario: This cop was on the periphery of the crime and either had witnessed or possessed information about the actual wrongdoers. Once identified, the feds would then go after the cop, apply their considerable legal muscle so that he suddenly faced a choice: Either get hurt by the government or cooperate. The pitch went something like this: We don't want to hurt you; we want your cooperation. But if you don't cooperate, we will hurt you. "This is how they work," one veteran Boston defense attorney said. "They target people, they give them bad choices, and they hope they choose Team America." In the end, the witness was a stepping-stone, a pawn in a larger prosecutorial game.

Paul Newman once starred in a film showcasing this classic federal strategy. The 1981 film, which also starred Sally Field as a newspaper reporter, was *Absence of Malice*. In it, a fictitious federal prosecutor named Elliott Rosen, based in Miami, is under intense pressure to solve the mob hit of a union boss. Rosen decides Newman, playing a liquor distributor and the son of a deceased crime figure, must know—or could find out—something. He puts the squeeze on Newman, having FBI agents tail him and IRS agents scour his books looking for something to pressure Newman into cooperating. Even though it's bogus, Rosen then leaks to reporter Sally Field that Newman is the subject of a federal investigation into the murder. Rosen hopes Newman, feeling his life might be in danger from others in the underworld, will come around. But Newman knows nothing—has, in short, nothing to trade—and he spends the movie working his way out from under the intense federal scrutiny and exacting revenge against the corrupt Rosen.

Ted Merritt was no Elliott Rosen. (Rosen's corrupt practices included having FBI agents illegally bug Newman's telephones.) But, like Rosen, Merritt was inheriting a big case in the unsolved beating of Mike Cox, a case where every other investigator had come up short. The Cox investigation was no different, either, from the standard operating procedure. Find a cop to squeeze. The question was: Which cop?

During the fall, Merritt and his investigators began combing through all the files and transcripts from prior probes. They had everything, including early reports filed to the police department's Internal Affairs Division, which had been off-limits to Bob Peabody. It was during this process of carefully reading their way into the case that they came across some of the material Peabody never had—namely, the incident reports originally filed on March 3, 1995, by Kenny Conley and Bobby Dwan. Merritt was fascinated by Conley's. In particular, Merritt focused on where Conley described chasing Smut Brown: "I then observed a black male, about 5' 8" in height, wearing a brown leather jacket, who I observed exit the suspect's motor vehicle from the right side, climb over a fence. With the suspect in view, I jumped over the fence and after a lengthy foot pursuit, I was able to apprehend the suspect in the rear of a building."

To Merritt, it surely felt like a discovery—a diamond in the rough. Here was a cop saying he scaled the fence where Cox was beaten. Conley must have seen something. It only made sense. Yet nearly two years had passed since Conley's report to Internal Affairs, and no one had pursued the matter further with him. Conley never even took credit for arresting Brown—that went to Richie Walker. It all seemed odd. Cops usually wrestled for credit. Was Conley hiding something?

It would be early 1997 before Merritt's investigation was going at full speed, but Kenny Conley had high potential. The Conley report vaulted to the top of the pile.

In his closing remarks to the Suffolk County jury on November 7, Bob Sheketoff said Lyle Jackson's shooting was "spontaneous" and not a planned killing, or joint venture, as the prosecutor was claiming. Focusing on the lack of credibility of the lone witness against Smut, he portrayed Marcello Holliday as an intoxicated patron at Walaikum's who couldn't tell up from down. "Could he have had three or four beers, maybe a few more than that?" He reminded jurors that while Holliday accurately described the others' garb, he got Smut's wrong. "He couldn't remember one single thing about my client, not one." Most important, he returned to Holliday's use of the word "whisper." "That place was abuzz." If Smut Brown said anything, how could Holliday have heard? "Did he say, 'That's one of them,' or did he say, 'I don't want any?'" Or maybe "'Sure, I'll get you a hamburger?'" Sheketoff said Holliday could not make out normal conversation, never mind a whisper. "Don't let some person who tried to tell you what a whisper was, whose testimony is contradicted by every other eyewitness in terms of the sequence of events, don't let him convince you beyond a reasonable doubt."

Sheketoff was feeling pretty good about his client's chances. He didn't think, however, it looked good for either Tiny or Marquis Evans. Both had taken the unusual step of testifying in their own defense. It was disastrous. Tiny's demeanor, hard looks, and stutter worked against him. He said his brother and Boogie-Down walked ahead of him into the restaurant. He lost sight of them, and, seconds later, the shooting began. Tiny's claim of complete ignorance was beyond belief. Marquis then testified his gun went off accidentally and just kept firing bullets on its own. Sheketoff, seated at the defense table, spotted a juror actually swivel in his chair to turn his back on Marquis. "Talk about someone who has stopped listening," Sheketoff thought. "How more clearly can you express you don't believe the witness?" In his closing, the prosecutor jumped all over Marquis's claim of accidental shooting. "Did you believe anything he said?" he asked. "He just happens to have the loaded nine millimeter handgun in his hand. He starts spraying the ground, because he's afraid. Is there a scintilla, a shred of truth to that? Even a shred? It's pathetic. That testimony was pathetic. Of course, that's for you to determine."

Jurors decided it was indeed pathetic. They convicted Tiny and Marquis of first-degree murder, and the brothers were sentenced to life in prison without parole. The jury then acquitted Smut and Boogie-Down. Smut was hugely relieved—he'd gotten justice, but he also considered it rough justice. He understood the legal theory of joint venture and thought if that fit Tiny and Marquis, it fit

Boogie-Down as well. But the government had nothing on Boogie-Down, and Boogie-Down was going home too.

Following the verdict, Smut turned in his seat and looked at his mother, his sisters, and Indira. His mother and Indira had come every day to the trial. Smut wore that signature smile of his—the curl that mixed mischief and deep relief. Ten minutes later he was a free man, embracing his mother in the hallway outside the courtroom. Then he swept up Indira. He thanked Bob Sheketoff. He had been behind bars for twenty-two months. The family rode the subway's red line to Mattapan and celebrated Smut's acquittal at Smut's old haunt. The last time Smut was in Conway's was the night Lyle was killed. Smut stayed for one drink; all he wanted was to go home and be with Indira and eight-year-old Shanae and five-year-old Robert Brown IV.

Sheketoff had something else on his mind. He'd just completed a murder trial with a bizarre side show—the unsolved Cox beating—featuring Boston cops testifying at cross purposes and contradicting one another. It stunk of a cover-up. Federal prosecutors, taking over the investigation, were looking to talk to Smut. Sheketoff was planning to encourage Smut to cooperate. In all his years, he'd never heard about such grotesque police misconduct and abuse of police—and, after two years, they seemed to be getting away with it. These cops, Sheketoff believed strongly, "needed scrutiny."

By early in 1997, Smut Brown found himself a part-time job in a manufacturing plant. He also resumed what he knew best—selling cocaine. In February, he was arrested by Boston police and charged with dealing coke and pot. He made bail. The FBI came around Mattie's house in Mattapan looking for him, but Smut wasn't home. Bob Sheketoff then called and explained Ted Merritt wanted him to testify before a federal grand jury investigating the Cox beating. But first, Sheketoff said, Merritt wanted to meet. "I was reluctant," Smut said. "I was telling him I didn't want to be in the middle of that. I was like, you know, I'm out here on the street and, you know, I don't want to be going against the police, man, because I'm nobody. Who's going to believe me?"

Sheketoff urged him to tell Merritt what he'd been telling him for a couple of years. Smut thought it over. He talked to his mother and Indira and decided to talk. "I thought it was right," Smut said. But he had other reasons for cooperating. The first had to do with his ill-will toward the Boston police. "In my own way it was like getting back at them for what they did to me. I'm not gonna sit here and say I wasn't bitter." Finally, there was Mike Cox. He respected Cox. "He always treated me fair."

The next month, he and Sheketoff showed up at Merritt's office. Besides Merritt, the lead investigator, an FBI agent named Kimberly McAllister was there. Smut told the feds what he saw in the seconds after he scaled the fence and before he took off on foot. He described Cox—who, at the time, Smut thought was Marquis—getting beaten. He identified Dave Williams as one of the beaters.

Then Smut said that standing next to the melee was a "tall, white guy."

Tall, white guy?

Yeah, Smut said. Tall, white guy—standing there by the beating.

Did he know the cop's name?

No, Smut said.

Ever see him before?

No, Smut said. Except he was the same cop who arrested me.

Merritt was eager to get Smut Brown before the grand jury.

Kenny Conley was half asleep in the living room of his second-floor apartment in South Boston. His tall, muscular frame was stretched out on the couch. It was Wednesday, March 26, raining outside, and Kenny was trying to get some rest after his overnight shift. There was a loud bang at the door.

He scrambled up off the couch. He'd been living in the second-floor walk-up at 720 East Seventh Street for only a couple of months. The triple-decker was owned by his twin sister Kris's boyfriend's brother. It sounded complicated, but that was Southie. Couldn't beat the friendly rental. Kenny was also feeling good about something else in his life—Jennifer Gay. He had been seeing Jen, pretty, petite, and brown-haired, since they met during Octoberfest on the waterfront. It was getting serious. For her work as a nursing home administrator, she'd moved during the winter to the town of Lee in western Massachusetts, but they were spending most weekends together in Boston.

Kenny pulled open the front door. The landing did not get much natural light and was a bit dark. Standing there were a woman and a man, both well-dressed.

The woman identified herself as Kimberly McAllister of the FBI.

We'd like to speak with you, she said. We think you can help us with a case we're investigating.

Sure, Kenny said. Come on in.

CHAPTER 15

The Perjury Trap

Five months later in New York City, the police assault of Haitian immigrant Abner Louima was the talk of the nation. Louima had been arrested following a disturbance outside a nightclub on August 9, 1997. Taken to the 70th Precinct station house in Brooklyn, he was beaten and sodomized in the bathroom by a patrol officer named Justin Volpe. Volpe kicked and pummeled Louima, and then, with Louima's hands cuffed behind his back, shoved a plunger up his rectum. By month's end, thousands of demonstrators were protesting outside city hall and the precinct station as part of a march called Day of Outrage Against Police Brutality and Harassment.

In Boston, when Kenny Conley's telephone rang on the evening of August 14, he took a deep breath. He picked up the receiver. It was his lawyer, one of two hired by the police union to represent him in a legal mudslide that had begun when the FBI agents came by his apartment.

His lawyer told him to sit down. Then she broke the news: Earlier in the day assistant United States attorney S. Theodore Merritt had obtained a three-count indictment charging Kenneth M. Conley with twice lying before his federal grand jury and with one count of obstructing justice.

Kenny, the indictment read, "did knowingly make false material declarations before the grand jury." In one count, the government charged Kenny was lying when he said he did not see Mike Cox at the dead end; in the second count, the government charged Kenny with lying when he said he did not see the beating. Both constituted the crime of perjury in violation of Title 18, United States Code, Section 1623. Then, in a third count, Kenny was charged with impeding "the due administration of justice, that is, a criminal civil rights investigation" of the Cox beating "by means of giving false, evasive and misleading testimony." This was in violation of Title 18, United States Code, Section 1503.

The eleven-page indictment, his lawyer explained, would be unsealed the next morning in U.S. District Court in Boston, where he would be arraigned. Kenny hung up the phone. He broke down. Then he called Jen at her place in western Massachusetts. She told him she was heading back to Boston and would be there as quickly as possible.

The next day, Kenny put on a suit and met his attorney at the Boston office of the FBI to officially "surrender." FBI agent Kimberly McAllister was waiting. She pulled out handcuffs and cuffed Kenny despite the protestations of his attorney, who pointed out that Kenny wasn't going anywhere and certainly wasn't a flight risk. Kenny was led away and his lawyer was told to find her own way to the federal courthouse.

The arraignment lasted only a few minutes. Kenny pleaded not guilty to the charges. He was released on a $10,000 bond and headed home. "I just felt alone," he said. For days he didn't want to leave the house. "I didn't eat; all I wanted to do was sleep."

If convicted, Kenny faced up to twenty years in prison and a $500,000 fine.

The two FBI agents had not believed a word of what Kenny had said in his living room. The agents had sat together on the couch where he'd been napping, and Kenny had begun the meeting by asking his guests if they wanted a cup of coffee.

No thanks, the agents had replied in unison.

Okay, what can I help you with?

Michael Cox.

What do you need?

Kenny, without hesitation, told them everything he knew—which wasn't much when it came to Mike Cox. The agents took notes, and later, in her three-page FBI report, Agent McAllister wrote up Kenny's account of arriving at Woodruff Way and chasing Smut Brown over the fence and into the woods. It was basically a rewrite of Kenny's original statement to Internal Affairs in March 1995 and of his interview with Jim Hussey. Nothing had changed. "He never observed any other individual on the fence or in that vicinity," the FBI agent had written. "Conley stated that he has no knowledge of who is responsible for Cox's injuries nor did he witness anyone assaulting him."

Within days of the interview, McAllister had called Kenny to tell him she was about to file her report with Ted Merritt. Was there anything he wanted to recant?

Recant? Kenny said. No, I have nothing to recant. Why?

The agent said, "I just think maybe you saw something you're not telling me."

"Which part is that?"

The agent said the part about not seeing anyone at the fence.

Kenny was confused. He'd been open and direct with the agents; he hadn't littered his answers with "I don't recall" or thrown up a fog of misdirection by saying, "Oh yeah, I did see people there, but nothing specific because it was all a blur." That wouldn't have been the truth, and it wasn't in him to be "cute," the way so many others had to avoid either further scrutiny or naming names. "I said what I said because I didn't see Michael Cox get assaulted. I didn't see him running after the suspect."

The agent asked one last time: Did he want to change anything?

"Ma'am, I don't wish to recant any part of my statement."

Ted Merritt seized on the agent's report as a gift. "He obviously didn't tell the truth when he was asked to," Merritt said later. Kenny even seemed to be mocking the federal investigators, expecting them to accept a statement they considered preposterous. In short order, Kenny Conley became the cop Merritt would target and convert into a cooperating witness against chief suspects Burgio, Williams, and Daley. Forget about Richie Walker or possibly Ian Daley—they were old news, so to speak, and the racial politics were far from ideal. Both were black, and how would it look for federal prosecutors to be squeezing black cops in the beating of another black cop? Forget other members of Mike's gang unit, such as Gary Ryan or Joe Teahan, given their association with Mike and the fact that their accounts, however conflicting, were so complicated.

Instead, Kenny emerged from the margins of the case to take center stage. He was a cleaner hit: the muscular white cop from "Southie," built like a brick, standing in the way of justice in the beating of a black cop. Right out of central casting for a prosecutorial drama set in Boston. Choosing Kenny meant that Merritt could exploit the neighborhood's image as the bigoted enclave violently opposed to busing to desegregate the city's schools. He could demonize "Southie loyalty" to explain why a cop like Kenny Conley would stonewall the feds to protect fellow officers. Look at Jimmy Burgio; he's from Southie too. Dave Williams? He and Conley were in the same class at the police academy. To Merritt, Kenny was the standard-bearer for the blue wall and its destructive and misguided basic principle: Never squeal on your own.

Merritt moved quickly following the FBI interview to get Kenny before the federal grand jury. The subpoena ordered him to appear May 29. Merritt called ahead of time to say he'd arranged to immunize Kenny with a court order saying Kenny would not be prosecuted for any information and statements he provided the grand jury. In theory, the grant of immunity was an investigatory tool to compel truthful testimony from witnesses reluctant to testify, often because of their own criminal liability. With this legal shield, the witness was free to testify honestly without fear of being charged afterward. The only exception was lying: Witnesses who lied faced perjury charges.

Kenny was planning to appear voluntarily before the grand jury. He didn't want any deals. No taking the Fifth. No grant of immunity. He thought witnesses who were immunized before testifying had something to hide; he wasn't like that. But no lawyer in his right mind allows a client to go before a criminal grand jury without immunity once it's been offered; it would be tantamount to legal malpractice. When he appeared on May 29 to testify, Kenny went along with the government's offer and was immunized.

Merritt had set a perjury trap. Kenny could either tell the grand jury the story Merritt believed Kenny was hiding—an eyewitness account of the Cox beating—or, if he insisted on sticking to that outlandish tale about not seeing Mike, then Merritt would see that Kenny was indicted for perjury. Merritt, of course, was hoping Kenny would fold right away. Then, riding Kenny's new testimony, Merritt could go after Mike's assailants. But if Kenny didn't, Merritt hoped he'd come around once criminal charges were filed. If not then, Merritt hoped he'd get it when the reality of going on trial hit him. Merritt would contact him periodically to let him know the case would vanish if he changed his tune. The prosecutor was a patient man; he'd keep up the pressure and wait Kenny out. Not one to second-guess himself, Merritt never wondered: What if I'm wrong about Kenny Conley?

Kenny felt like beating his head against the wall. "They thought with the immunity I was going to say to myself, Okay, I'm in the clear now. They thought I was going to go in there and tell them I saw Jimmy Burgio do this, I saw David Williams do that." But Kenny couldn't do that. Indeed, questioned by Merritt and now under oath before the grand jury, he once again described Smut Brown running from the Lexus.

"You saw him go up the fence and get down on the other side?"

"Yes," Kenny said.

"And where did you make these observations from?"

"As I was running towards him."

Merritt then loaded up and asked Kenny a set of rapid-fire questions about Mike Cox: Did you see "another individual" running after Smut Brown? Did you see "anyone else in plainclothes" right behind Brown? Did you see an officer in plainclothes at the fence "standing there, trying to grab" Brown as the suspect scrambled over the fence?

To every question, Kenny said: "No, I did not." Openly and politely, he answered everything Merritt threw at him—to a fault. Merritt, finishing up, masterfully baited Kenny with a hypothetical question incorporating information he was gathering from others about the events at the fence: "If these other things that I've been describing, a second—another plainclothes officer chasing him and actually grabbing him as he went to the top of the fence, you would have seen that if it happened?"

Kenny, without hesitation, said, "I think I would have seen that."

It was the kind of exchange that makes any prosecutor's list of greatest moments. A more sophisticated witness would know to avoid this sort of hypothetical question that was like a bed of leaves concealing a steel trap. But Kenny was being Kenny—ever-helpful, straightforward, and even naive. Yeah, if Mike Cox was hard on Brown's heels and grabbing at Brown at the fence, *I think I would have seen that.*

Kenny hadn't seen Mike, but Merritt, in that quick repartee, had finessed it so that Kenny had concurred with the very argument Merritt was planning to use to prosecute him: Kenny Conley himself says he would have seen Mike Cox if Cox and Brown were at the fence together; ergo, dear jury, the fact that Conley denies seeing Cox means Conley is lying! Case closed.

Next to losing his mother, the federal indictment on perjury charges was the worst day in Kenny's life. He was immediately suspended without pay from the force. The department seized his badge and gun. But while he tumbled into an emotional free-fall, his two sisters and close friends rallied the only way they knew how.

The tickets—about the size of a business card—began selling early in November. Poster-sized versions appeared in storefronts around the neighborhood, one-by-two-foot slabs of cardboard announcing, "A Time for a Friend." The party was scheduled for Friday night, November 28, nearly three years to the day from the death of Kenny's mother. The ticket promised a "DJ" and "Raffles" and requested a donation. Kenny first learned about the fund-raiser when he spotted a poster in Java House, the shop on East Broadway around the corner from his family's house on H Street, where most mornings he grabbed a light coffee. Then, as the end of November approached, a flier circulated in his old district station, where Kenny was now officially persona non grata. "Let's Go Folks!" the flier began. "Ken Conley Time." The date was included, along with the names of the contact persons: "McDonald, Mags or Hopkins." Danny McDonald had been Kenny's partner, except for the night he rode with Bobby Dwan. Billy Malaguti was a veteran cop and union representative. Tommy Hopkins was one of Kenny's pals.

Kenny had specifically told his twin sister, Kris; his sister Cheryl and her husband, John; as well as his boyhood pal and fellow cop Mike Doyle, that he did not want a party. He didn't want to be the center of attention; he was ashamed about the fix he was in and just wanted to hide. But that was like asking them to deny their roots. Holding a "time" for a friend was the Southie way; it was in their DNA.

The night after Thanksgiving, a clear, cold night with temperatures below freezing, hundreds of Kenny's family and friends poured into the hall rented for the occasion. The Florian Hall in Dorchester was a Boston fixture, a single-story brick building famous for blue-collar parties of all kind—family, political, and work-related. It was visible from the southbound lane of Interstate 93 running through the city. Highway drivers that night not only caught a glimpse of a sea of parked cars, including police cruisers, and partygoers streaming through the double front doors under the hall's own red-neon sign, they also saw Kenny's name flashing on the digital billboard erected alongside the highway at the headquarters of Local 103 of the International Brotherhood of Electrical Workers.

Besides friends, plenty of off-duty cops were in attendance. Bobby Dwan, for one, was there. So too were a few sergeants who were Kenny's patrol supervisors. Cops who knew Kenny and, like Bobby Dwan, knew Kenny was a straight-shooter. But Kenny also realized there were cops showing up, cops he didn't know very well, who assumed he *was* lying about Woodruff Way to protect fellow officers. Kenny got that feeling when some came over to offer a thank-you with a knowing wink and a nod. But, said Kenny, they were wrong to think that about his grand jury testimony. "I was telling the truth. I was not going to risk my career and family. I was just getting started with Jen."

It led to awkward moments. Kenny was standing in the hall drinking a beer, talking with friends, when a hand came up and over his back from the crowd behind him. Kenny grabbed the hand. "I didn't know who the hell I was shaking hands with," he said. He turned to look and froze. "It was him." Jimmy Burgio was pumping his hand. Kenny walked away, "pissed off, I would say." And Burgio wasn't the only unwelcome guest. Dave Williams showed up, but as soon as word got around among Kenny's friends, he was asked to leave. The cold shoulder was hardly the stuff of a one-for-all-and-all-for-one.

Kenny would go back and forth in the months afterward on whether the party was a lousy idea. Celebrating a cop accused of perjury and obstruction of justice certainly looked bad. But he could not fault what was at its core—family support. "Maybe it was bad timing, but it was out of the kindness of these people, getting together. No one was thinking about the politics of it." Kenny couldn't worry about appearances. Besides, the party did serve its purpose. It buoyed his spirits, knowing his family and his many friends from the old neighborhood believed him, even if federal prosecutors, the FBI, and the Boston police brass did not. And over time, the grassroots support for Kenny continued to spread. Stickers appeared on car bumpers, and signs were propped in the windows of homes around Southie: "Justice for Kenny Conley."

But no matter what Kenny believed in his heart, the "time"— a standing-room-only affair that ran deep into the night—proved a public relations disaster. The party played right into Merritt's theory that Kenny was standing tall for his brothers in blue; now fellow cops were rewarding him by turning out in a show of support. The media bought into this notion of the party as a symbol for the sinister side of police solidarity. Tipped off, a team of *Boston Globe* reporters staked out the hall, and, as part of what amounted to the first in-depth report by the city's media about the unsolved Cox beating, the newspaper ran a photograph artfully capturing a police cruiser and a throng of men entering the hall. The story, examining the failure to solve the assault, was headlined: "Boston Police Turn Against One of Their Own: Years After Beating, Officer Has Seen No Help from Colleagues." Echoing Merritt, the message was clear: If Kenny Conley agreed that he had been in a position to see Cox but insisted he didn't, then Conley must be lying.

Once Mike's lawyer, Steve Roach, caught wind of Merritt's new twist in the investigation, he added Kenny Conley as a defendant to Mike's lawsuit, joining Jimmy Burgio, Dave Williams, and Ian Daley as the officers cited specifically as violating Mike's civil rights. The lawsuit accused Kenny of essentially the same wrongs as Merritt's criminal charges—of covering up the beating by lying to the grand jury and failing to cooperate with investigators. Mike had no idea if Kenny Conley was lying. He didn't know Kenny, beyond the fact that they were about the same age and had grown up in Boston. But Mike understood why Steve Roach wanted to go after him. The logic of Merritt's point of view was seductive, and Mike certainly knew cops often practiced the art of omission—as if using Wite-Out to erase parts of their memory—to avoid implicating another cop.

While his lawyers worked on his case, Mike worked on holding his world together. His resolve to get to the bottom of the attack didn't change the mixed bag of setbacks and successes making up the day-to-day. He took the examination for promotion to lieutenant in late 1996. It was like the Mike Cox of old—the one with a once-and-future distinguished police career. But Mike flunked the test, a first for him. His concentration wasn't the same; he'd robotically gone through the motions. The next year, though, he achieved a personal milestone when at long last he finished Providence College, earning a bachelor of science degree in business management. He and Kimberly were also able to enjoy Craig Jones's wedding, where Mike served as best man, and then, in the spring of 1997, the Coxes celebrated their own good news: Kimberly was pregnant. In January 1998, they had their first girl, a baby they named Mikaela, joining nine-year-old Mike Jr. and eight-year-old Nick.

But all around there seemed to be reminders of the beating. Commissioner Paul Evans announced a new round of personnel changes in late 1996 that, for Mike, amounted to another bizarre and confusing signal about Evans's supposed promise to get to the bottom of the beating. Evans transferred Sergeant David C. Murphy to the Internal Affairs Division. He then promoted him to sergeant detective. Murphy was joining Dovidio in the department's Bureau of Internal Investigations—two sergeants who were part of the supervisory failure that saw the cover-up at Woodruff Way take off and flourish. Mike did his best to avoid them, but there they were, all in the same bureau.

Mike saw land mines inside his division and out. He sidestepped them as best he could. He would work police details while off duty to make extra money. But in April 1997, when Mike was assigned to supervise a contingent of officers from the C–11 station in Dorchester as part of the 101st Boston Marathon's extensive public safety coverage, he backed out. Mike didn't want to risk running into two of C–11's finest, Jimmy Burgio and Dave Williams.

"Virtually everyone that I work with, it seems like, is involved in some way, shape, or form with this case," Mike said, "whether they are friends with somebody who has been named, or whether their supervisor could be named." The department "can be a very small place," he said, and Mike saw trouble everywhere. "I don't trust anyone."

Mike and Kimberly had moved from their apartment on Supple Road into their first home. The couple paid $145,000 for the three-story, wood-shingled colonial on Rundel Park in August 1996, fixed it up, and moved in the next summer while Kimberly was pregnant. The six-bedroom house in Dorchester, on a short dead end atop a gentle slope that gave it a commanding presence, featured a picket fence, a gabled roof, and a large front porch. The house enjoyed the shade of two towering trees in the large side yard. But it wasn't as if the relocation provided Mike any refuge from the storm. The harassment kept coming, despite a new and unlisted telephone number.

The nightmares continued as well, some seeming to come to life. Late one night the first summer in the new house, the doorbell awakened him. Mike climbed out of bed and reached for his Glock semiautomatic pistol. He tiptoed down the stairs. The front windows were open to the breeze, and through them Mike heard the crackle of voices on a radio. In one window he saw the clear silhouette of a man in uniform standing by the door. Mike thought he recognized the officer. Mike stood still. The doorbell rang again. Mike chambered a round. The weapon wouldn't do him any good unless it was loaded. But he mainly wanted to signal the unexpected visitor that he was armed, and he purposely made plenty of noise while pulling the gun's chamber.

"You can put that away," a voice said from outside.

Mike turned on the porch light. He was right; it was Dave Williams.

Mike opened the door, his pistol at his side.

"Someone broke into your car."

Mike looked past Williams. He saw the rear door of the car parked on the street was open. The interior light was on. The Coxes now lived in a neighborhood in the Area C police district, and Williams was apparently working his usual overnight shift.

Williams asked Mike to see if anything was stolen. Mike stepped out onto the porch. "No," he said, "I don't think so." Williams asked Mike again to check the car. The two men exchanged looks, the eyes of two tigers sizing each other up.

Mike walked to his car, looked around, and firmly shut the door.

"No," he said. "Nothing's missing."

"Okay," Williams said.

They passed each other. Williams climbed into his cruiser. Then he drove away.

For Mike, it was another of his night's many mysteries. He had no idea what to make of it. He simply climbed the stairs and went back inside, where he resumed the nocturnal ritual of tossing and turning and worrying about his family's safety.

CHAPTER 16

A Federal Miscarriage of Justice

When the police union hired Willie J. Davis and his partner to represent Kenny against Merritt's investigation, it didn't take Davis long to understand Kenny's predicament. Davis was a veteran of courtroom battles going back more than two decades, first as a prosecutor and then as a defense attorney. In the mid-1950s, the Georgia native had been a standout in the backfield of the Morehouse College football team, the college in Atlanta where, in 1987, a visiting student named Mike Cox met a Spelman College junior, Kimberly Nabauns. By the late 1950s, he'd moved north to Boston, where he attended law school and was named an assistant state attorney general by the state's first black attorney general, Edward W. Brooke III. He went on to become an assistant U.S. attorney and then was the first black to serve as a U.S. magistrate on the federal bench in Boston. He entered private practice for good in the mid-1970s. He'd become friends along the way with another young state prosecutor, George V. Higgins, the future novelist known for gritty Boston-based crime stories, especially *The Friends of Eddie Coyle*. When he left government work in 1976, Davis joined Higgins in representing former Black Panther Eldridge Cleaver in California.

It was in 1970, while still a federal prosecutor, when Davis had what he'd later refer to as an experience matching Kenny Conley's.

Working with federal drug agents, Davis secured the indictment of two Boston men on cocaine trafficking charges. The two had long records for violent crime and were widely feared. The prior year, they'd beaten a murder rap in the shooting deaths of a Roxbury civil rights leader and two others. Police and federal agents were looking to arrest them. The next day, Davis and an agent, driving away from a meeting in Roxbury, overheard on their radio that one of the wanted men had been spotted nearby at Dudley Square. Dennis Chandler, known as Deak, was with an unidentified friend. Agents were moving in. Davis was just a block away. He and his companion headed to Dudley Square to join the capture. They pulled up alongside other law enforcement vehicles. The prosecutor had no business getting involved in an arrest, but he was caught up in the moment. "We moved in," Davis recalled later, "and everyone had a gun but me." Davis recognized Chandler and locked in on him. "I was looking right at Deak because I knew Deak was bad."

But as Davis ran he missed the other man with Chandler. Then he missed seeing the other man draw a handgun. And he missed the man raise his arm to point the gun at him. "I never saw any of this," Davis said. In the next instant, a federal agent, running from the side, jumped the gunman. The pistol was knocked loose and clanked along the sidewalk. "It was the first time I realized there was a second person there," Davis said. "I was concentrating so hard on Deak, I didn't see."

For Davis, whether he believed Kenny Conley or not, defending the client was a professional responsibility. But this was an instance where the lawyer fully believed the client. And, by coincidence, while Davis was taking up Kenny's defense in 1997, what Davis knew firsthand as "tunnel vision" was being called something else in the research laboratories in the psychology departments at Harvard University in Cambridge, Massachusetts, and at Kent State University in Kent, Ohio. Daniel J. Simons, a Harvard psychology professor, was conducting experiments into "inattentional blindness" or "selective attention" on the eighth floor of the university's William James Hall. In the argot of the academy, the phenomenon of inattentional blindness was defined as this: "When people attend to objects or events in a visual display they often fail to notice an additional, unexpected, but fully visible object or event in the same display." In lay terms, Professor Simons explained during an interview, "People count on something unexpected to grab their attention. But often this doesn't happen. The consequences can be dramatic. Like people assuming a car pulling out in front of them will catch their attention, but it doesn't, and there might be an accident." It was in the context of car safety and cell phone use that Simons's work eventually moved from obscure journals into the public domain. The writer Malcolm Gladwell wrote extensively about inattentional blindness in an article, "Wrong Turn: How the Fight to Make America's Highways Safer Went off Course," published in the *New Yorker* magazine in June 2001.

In 1998, Simons and his colleague Christopher Chabris conducted perhaps the most astonishing experiment—what they playfully called "gorillas in our midst." The researchers began by making a videotape showing two teams of three players moving in a small open area. One team wore white T-shirts and the other team wore black T-shirts. Each team had a basketball and passed it only to teammates. The video lasted about a minute.

Observers were brought in to watch the videotape played on a television monitor. Their task was to focus on the white team and to count the number of its passes. Then came an unexpected event none of the observers were told about beforehand: About halfway through the action, a researcher wearing a full-length gorilla costume walked into the scene and stopped in the middle of all the ball passers. The gorilla turned to face the camera and beat its chest a few times. Then it turned and walked off camera. Nearly half of the observers *missed* the gorilla. The dramatically high rate of "blindness," Simons noted, showed that "when people are engaged in an attention-demanding task—doing something that requires their attention to be focused on some parts of the world and not others—often they do not see something that is very visible, very salient, but unexpected.

"This doesn't match at all with people's intuition."

It was that "people's intuition" that formed the core of Merritt's conviction that Kenny Conley was lying. Even Kenny acknowledged that—saying he should have seen Mike at the fence. The new experiments and data at Harvard quite possibly offered an explanation for why he hadn't. But Kenny and Willie Davis were unaware of the scientific studies under way several miles away across the Charles River. Kenny would continue to talk in a vague, general way about tunnel vision. The conflict was that he didn't know why he hadn't seen Mike Cox, and Ted Merritt thought he had. Merritt kept contacting Willie Davis periodically to say if Kenny would come around his troubles would end. It drove Kenny to existential despair. "What's going on!" he'd tell his lawyers. "I'm not going to lie for the government. I'm not going to lie for Burgio, Williams, or Daley. I'm not going to lie. That's it." He even wished he had seen Mike. "I wish I could turn around and tell Ted Merritt, Yeah, I saw him pulled from the fence. But I can't."

The trial was fast approaching. He and Jen talked about the case. Kenny wanted her to grasp the possibility he might be convicted. Jen wasn't surprised that Kenny wouldn't lie, but she wondered, like everyone, why he hadn't seen Mike Cox. But she never doubted him. She certainly didn't want him to go and say what they wanted to make the case go away. "Just say what you know," she told him. "How can you get in trouble for that?"

In the case he built, Merritt had no direct evidence Kenny was lying. He did not have a witness who could testify he actually saw Kenny looking right at Mike Cox or at the beating. Merritt's case was circumstantial, built mainly on three legs—what he would call the "interlocking" testimony of three witnesses: Mike Cox, Richie Walker, and Smut Brown. Mike said he was right behind Smut Brown during the short run to the fence. Richie Walker seemed to back that up, saying he saw Mike chase Smut Brown to the fence. Smut Brown then finished it off, saying once over the fence, he turned and saw a "tall, white guy" standing by the assault of Mike

Cox—the same white cop, Brown also surmised, who eventually captured him after a foot chase through the woods.

It was a tightly woven package. Except for one problem—Merritt's case was deeply flawed. For one, Richie Walker had become a mess of contradictions and unreliability. The prosecutor and FBI agent McAllister, studying Walker's prior statements, noticed that Walker had changed his story. He'd told Internal Affairs he saw another cop running behind Mike, but he never mentioned a word of that during Bob Peabody's investigation. The FBI agent asked Walker about the discrepancy. Walker came up with a lame-sounding explanation. The first account was untrue, he said, a tale he concocted because he was trying to help. "He felt this way because he knows victim and likes victim," the agent wrote afterward. "He felt bad that he could not say what happened and therefore convinced himself that he actually saw someone or something."

The agent had asked Walker to take a polygraph test, and Walker said okay. Then Walker, apparently hopelessly confused, proposed his own truth-seeking exercise: hypnosis. It was a zany, almost circus-sounding idea. Not surprisingly, the FBI did not take him up on it. But Walker didn't take a polygraph either; he balked the day of the test, saying he'd changed his mind.

The muddle was all there in the agent's typewritten, two-page FBI report—a document the government was required to turn over to Willie Davis during the pretrial discovery phase. The legal nickname for the exculpatory information was "Brady material," named after the 1963 case *Brady v. Maryland*, where the U.S. Supreme Court made crystal clear the constitutional importance of the rule: "The suppression by the prosecution of evidence favorable to an accused upon request violates due process."

But the government never turned over the radioactive memo. In one letter from Merritt that included a section marked "Brady material," where the memo should have been cited, the letter simply said: "None." Without the Walker memo, Willie Davis would never get the chance to shred the credibility of a key government witness.

Then there was Smut Brown. Smut appeared before the grand

jury the month following his meeting at Merritt's office in the federal courthouse. He told the grand jury he saw Dave Williams whack a man at the fence. The man was then beaten by officers, he testified, both black and white, dressed in uniform and in plainclothes.

Merritt then asked Smut questions crucial to the pursuit of Kenny Conley. Smut once again said he saw a tall, white cop in plainclothes standing near the beating and that later, after a foot chase, he was caught "by the big white officer."

With that, Merritt got what he wanted—an apparent identification of Kenny at the fence. Kenny had captured Smut—nobody questioned that—so if Smut was saying the white cop who ran him down was the same white cop he saw at the fence, then Smut Brown was saying Kenny Conley was at the fence. It was a looping but compelling deduction. Merritt left it at that. He didn't question Smut any further about why Smut thought the cop who arrested him was the same cop he'd seen at the fence. He never had Smut try to pick Kenny Conley out of a photo array or a lineup.

Why risk it? Merritt was looking foremost to squeeze Kenny Conley. He'd rather have Kenny fold than actually take him to trial. But if Kenny was determined to stonewall, Merritt was going all-out, a full-court press. For that, Merritt had secured from Smut the evidence he needed, a circular but nonetheless compelling inference that sounded like Smut was directly fingering Kenny. There was "no prosecutorial purpose" to trying to verify Smut's account with a photo array. It was useful as it was: a virtual identification. "For the purposes of the case, that was sufficient," Merritt said later.

Seeking the truth apparently wasn't deemed necessary.

Smut Brown warily made his way down the third-floor corridor of the federal courthouse, flanked by Indira and his mother, Mattie. They were looking for Courtroom 11. Finding it, they sat on a wooden bench outside. It was Friday, June 5, 1998, the fourth day in the perjury trial of the *United States of America v. Kenneth M. Conley*. Outside, the midmorning weather was comfortable, with partly cloudy skies and temperatures in the high 60s. Inside, the windowless hallway was stuffy and stale. With a new $225 million U.S. District Courthouse due to open in September on the South Boston waterfront, majestically overlooking Boston Harbor, the government wasn't much interested in the upkeep of the old and worn-out, landlocked facility.

Smut looked around, sizing up the other people hanging out in the hallway. Friend or foe? He was outside his turf; it was as if he'd wandered into a Roxbury or Mattapan neighborhood where he wasn't welcome, controlled by a street gang he had no ties to.

He'd promised to testify, but that didn't mean he was happy about it. He was convinced Boston cops were after him for cooperating in the Cox case. Soon after testifying at Merritt's grand jury, he was busted for selling coke at a Sunoco gas station in Dorchester. He'd driven his red Nissan Maxima to meet a buyer who'd beeped him. Unfortunately for Smut, cops were conducting a stakeout. Initially they thought Smut was his younger brother. Smut was taken in, booked, and put in a holding cell. That's when things got screwy. Figuring out who he was, one of the arresting officers confronted him. The officer acted as if discovering he'd busted Smut Brown had made his day. "He said to me, 'You ain't nothin' but a piece of shit,'" Smut said. The cop then stuck his hand in the rear pocket of Smut's pants and, after fiddling around, magically pulled out a plastic bag of crack cocaine. Oh, look what we have here! More evidence!

Eventually, Bob Sheketoff exposed the fact that police had planted the drugs on his client. During the suppression hearing in court, the lawyer vented his frustration. "This case does make me want to make speeches," he told the judge. "Twenty Boston police officers at the scene of an 'attempted murder' of another Boston police officer, and the only one that will speak up and say anything about it is a drug-dealing, car-stealing citizen who, the second he speaks up and says anything about it, finds himself with one problem after another with the Boston police department." The judge brushed aside the lawyer's rant, saying it was unnecessary for him to decide the motive for the police misconduct and that the wrongdoing was what mattered. The judge ruled the crack cocaine was inadmissible, and, with that, the case against Smut fizzled.

Smut was left thinking, however, that he wore a bull's-eye on his back. He sat in the courthouse, slated to be the third witness called by prosecutor Ted Merritt. The courtroom was packed with reporters and spectators, including Jen, Kenny's sister Kristine, and Kenny's friends from Southie and from the police force. Like Kenny, they tended to be big and fit-looking, and shoulder-to-shoulder in the courtroom, they resembled a defensive line on a football team. Or, as Merritt would have it, a blue wall. The first witness was Mike Cox, who was in court with Kimberly. Merritt walked Mike through what he remembered about running to the fence after Smut Brown.

Federal prosecutors then called Richie Walker, the second leg in his interlocking web of circumstantial evidence against Kenny. On the stand, Walker displayed none of the shakiness reflected in the explosive FBI report that sat in government files rather than where it should have been—in Willie Davis's hands. Walker completed his testimony by describing seeing Mike reaching for Smut Brown at the top of the fence.

"Members of the jury," the judge then said, "we'll take a ten-minute recess."

The doors flew open as spectators cleared the courtroom. FBI case agent Kimberly McAllister came out and walked over to where Smut was seated with his mother and Indira; it was part of her job at the trial to "babysit" the government witnesses. Smut was immediately put on edge by the surge of off-duty cops and other bulked-up white guys spilling out into the hallway. A man in the corridor suddenly caught his attention. His heart jumped. He turned quickly to Mattie and Indira, his voice cracking. "That's him," he said. Mattie asked what he was talking about. Smut, pointing out a tall, white guy walking away from them, said, "Him! The guy at the fence!"

Smut turned and told the FBI agent—there's the cop he saw at the fence. He did not know the cop's name but said, that's him. "She's like, 'No, no, no, no, that's not the guy,'" Smut said. "I'm like, 'Yo, it is the guy! I know it's him.'" It would take Smut a while to sort through the moment. He'd always thought the cop he saw at the fence was the same cop who'd arrested him. He thought that because he'd gotten a solid look at the white cop at the fence—he was tall, white, and wearing a cap. Flat on his stomach a few minutes later, he barely saw the cop who captured him, but from a glimpse he *assumed* he was the cop from the fence. "He's big, too, with a hat on, and I'm thinking—same guy." Smut had never had reason to doubt the assumption. Ted Merritt had certainly never asked Smut to explain why he thought the two cops were the same person. Left unexamined, it seemed as if Smut was saying he'd seen Kenny Conley at the fence—a looping deduction that was good for Merritt's case.

But Smut was realizing this was all wrong. He'd incorrectly merged two cops into one. The mistake meant that walking down the corridor and out of sight was a potential lead in the Cox investigation. Smut tried explaining this to the FBI agent, but she didn't seem interested. The agent was working to settle her witness down. The short recess was ending, and she wanted to get him on the witness stand.

Smut entered the courtroom feeling flattened. He looked over at the defense table at Kenny Conley. "I have no clue who he is," he thought. "They got the wrong guy." Smut walked to the witness stand. He was beginning to sense he was being used.

"Now, please speak into the microphone," Merritt said after he was sworn. "State your name."

"Robert Brown."

"How old are you, Mr. Brown?"

"Twenty-six."

Smut kept waiting for the courtroom finale when the witness is asked to point out a defendant: Can you please point out the tall, white guy you saw that night at the fence?

No, Smut could not. "I would have said, 'I seen him out in the hallway.'"

But the Perry Mason moment never happened. Merritt had never needed Smut to directly identify Kenny Conley before, and he wouldn't now. For his part, Smut was not about to interrupt the proceedings. "I just figured, man, I just want to get this over with. Go get the hell out of there." Instead, Merritt artfully unveiled the powerful inference created by Smut's statements that made it seem as if Smut saw Kenny at the fence.

Everyone came away thinking a clear identification had occurred, including Kenny and his attorneys. For his cross-examination, Willie Davis took the standard tack and attacked Smut as a coke-dealing lowlife whose testimony was unreliable. Merritt's legal sleight of hand worked so well, in fact, that the next day's news story in the *Boston Globe* drew the following conclusions from Smut's testimony: "Boston police Officer Kenneth M. Conley stood nearby as a fellow officer was beaten senseless by three or four other officers who mistook him as a suspect." It had sure sounded that way, even though Smut never said Kenny Conley and always said the "tall, white guy." Brown, the newspaper reported, "said he made momentary eye contact with Conley, who chased, captured, and arrested him."

Kenny badly wanted to take the stand in his own defense, but having a defendant testify was always high-risk. His lawyer, Willie Davis, convincingly argued against it. Instead, Davis tried his best to attack the credibility of Merritt's case and argue that with the chaos of the dead end and Kenny's "tunnel vision," it was reasonable to believe Kenny missed seeing Mike Cox or the beating while chasing Smut Brown. He asked rhetorically, if Kenny were lying, why would he say he ran to the fence? "Why would he put himself there?" If Kenny wanted to lie and cover up, he said, why didn't he concoct a story about being far away from the beating? "Why didn't he do that?"

But by the time Willie Davis made his closing argument to the jury, little had gone well. Even little things—such as courtroom atmospherics—worked against him. One juror complained to the judge she felt intimidated by Kenny's wall of friends. The judge ordered the spectators to vacate the front row and sit farther back. More important, Davis was at a disadvantage. He was without the explosive FBI report on Walker, lacked the knowledge and resources to call experts about "inattentional blindness," and, like everyone else, missed how Smut's virtual identification had been manipulated by the assistant U.S. attorney.

Merritt was the one with all the cards and, in closing, he expertly argued Kenny Conley, in defiance of common sense, had lied to protect Jimmy Burgio, who Kenny knew "growing up in South Boston." Referring to Dave Williams, Merritt inserted the distinguishing characteristic, "Kenneth Conley's academy classmate." He recapped key testimony and reminded jurors that Smut Brown watched officers beat Mike. "Brown then saw a tall, white plainclothes officer around that group," he said, "and Brown took off, eventually being captured by that same tall, white plainclothes officer, who you know is the defendant, Kenneth Conley." It wasn't tunnel vision that prevented Kenny from seeing Mike, he argued. "It was a deliberate cropping Cox out of the picture."

Finally, in a flourish, Merritt made clear the motive for Kenny's stonewalling: the blue wall of silence. "When a witness takes an oath to tell the truth in the grand jury, there is no exception for police officers who don't want to implicate another police officer who violated the law." He then noted the power of the code, emphasizing Kenny chose to lie even when the beating victim turned out to be another police officer. "What does that tell you about the power, the forces that were motivating Officer Conley, and what does that tell you about the chances when a victim is a citizen?"

Seated next to his attorney, even Kenny was impressed by how the prosecutor played to the jury's passions. "Listening to Ted Merritt, I think I'm guilty," he said later.

The jury thought so too. Following seven hours of deliberations, the jury found Kenny guilty of one count of perjury—of saying he'd not seen Mike Cox at the fence—and one count of obstruction of justice. When the verdict was announced, Kenny, dressed in a brown suit, began rubbing his forehead hard, as if trying to comprehend the news. He hung his head and slumped over the table. Behind him, Jen, his sisters Kris and Cheryl, other relatives, and friends let out gasps. Kris, not one to cry easily, burst into tears. It was hard for her to listen to the judge address Kenny "like the scummiest of criminals.

"Kenny was no longer this decent guy with tons of potential," she said. "He was found guilty and had to put his affairs in order to go to prison. It was mind-blowing."

One veteran police officer afterward called the verdict a "lose-lose" situation. "Everybody feels sorry for Cox," he said. "But Conley is just a pawn being played."

Three months later, on September 29, Kenny was sentenced to serve thirty-four months in a federal prison and fined $6,000. Kenny stood to speak. "I have felt bad for Michael Cox," he told the judge. "If I could help him I would have—if I knew who did it."

Merritt had won the case, but the victory proved pyrrhic. He kept after Kenny to change his story, but he would not—could not. Kenny instead began fighting for his name, appealing the conviction and gaining supporters. It was all part of the gross miscalculation on Merritt's part—to devote a year or more in the pursuit of Kenny Conley in the mistaken belief that he was the witness who could break open the Cox beating. Merritt's criminal investigation, though still ongoing, became stymied and stuck in the morass of the fallout over Conley's conviction. Instead of opening doors, the Conley matter became a bitter and paralyzing distraction.

Three years had passed since the night Mike was attacked and abandoned on Woodruff Way, and none of the investigations had gotten to the truth. First the police department's Internal Affairs Division came up short. Then Bob Peabody's Suffolk County grand jury investigation faltered. Finally, Merritt and his team of federal investigators not only failed, they'd undermined justice with the wrongful conviction of Kenny Conley. It was the investigatory equivalent of three strikes.

Mike Cox was left to find his own justice. His was the last case standing.

CHAPTER 17

On His Own

Three days after Kenny Conley was sentenced to prison, Jimmy Burgio, burly and barrel-chested, was working a paid detail at Nancy Whiskey's, one of Southie's more rough-and-tumble bars. He was stationed at the door in his Boston police uniform.

It was Friday, October 2, a cool night in early autumn. Burgio had a regular gig at Nancy Whiskey's. He liked the extra money, of course, but he also liked the body contact. The bar drew a hard-drinking crowd; on occasion, a melee would erupt, like a bench-clearing brawl in ice hockey, the sport Burgio was fanatical about.

Burgio's life had been spinning out of control. He was the target of Ted Merritt's federal criminal investigation into the Mike Cox beating. He was facing trial in December when Cox's own civil rights case against him, Dave Williams, Ian Daley, and Kenny Conley was scheduled to begin in federal court. And just the past June, he had been accused again of police brutality. It had happened at Nancy Whiskey's after the two o'clock closing, when a firm hand was often needed to clear out the barflies. Burgio had gone inside to assist two bouncers remove a recalcitrant patron. The guy agreed to leave, but said in a minute. Moving with hurricane force, Burgio grabbed him by the arms, twisted them behind his back, and pushed him out the front door.

To Burgio, the moment itself was unremarkable. But a letter followed and, with it, the reason that the incident had stayed with him. The man's lawyer wrote Police Commissioner Paul Evans to say they were going to sue Burgio, the police department, and the city. The patron alleged Burgio slammed his head into the door, punched him until he was bloodied and unconscious, and then threw him outside onto the sidewalk. "It is my client's position that Officer James Burgio used excessive force," the letter said, "and that the Boston Police Department and Mayor Menino was negligent in the supervision, discipline, control and training of its officers."

Like Jimmy Burgio needed another headache. The city eventually paid the man $86,250 to settle his claim, but Burgio, as always, denied any wrongdoing; he was just doing his job to keep the city safe.

Burgio was seated outside Nancy Whiskey's—near closing time, again—when a car drove up and parked, and a bunch of people climbed out, including Kenny Conley.

Kenny spotted Burgio right away. His friends did too.

"There's Jimmy," one cautioned. "You want to go in there?"

The group had come from a fund-raiser for the Special Olympics at a union hall in Southie. No one had known Burgio was working the door. Kenny had what he called a "package on," meaning, "I was drinking." Since his sentencing, he'd been drinking hard—too hard.

Maybe they should go somewhere else, another friend suggested.

Kenny looked at Burgio dressed in his uniform—a reminder that throughout the Cox investigation, Burgio had stayed on the street, a full-fledged member of the police department. Kenny lost his badge the year before on the day he was indicted.

Kenny pushed the car door open. He'd made up his mind. "Fuck him. I'm not letting him keep me out of a place where I live."

He walked past Burgio and into the bar. He and his friends ordered a round, but Kenny didn't have much to say. Unless things changed, he was going to prison. In court earlier in the week, Ted

Merritt had sought a sentence of forty-six months, a year longer than the thirty-four months the judge ended up imposing. Even with the lesser amount of time, his lawyer, Willie Davis, was demonstrably upset; by comparison, he noted that Stacey Koons, one of the L.A. cops caught on videotape beating Rodney King, had received a thirty-month term. "Tell me that's justice," Davis had told reporters while shaking his head. Kenny's twin sister, Kris, was hurrying her wedding plans to make certain Kenny could attend. Kenny might have been out on bail and appealing his conviction, but he had nightmares about being scooped up off the street and taken away to prison.

He stood inside the bar thinking about Burgio. "He's sitting there working like nothing's going on, nothing's happened. Three days prior I had just been sentenced to thirty-four months for something I had nothing to do with." It was starting to drive him a little crazy. He and his friends drank a few beers, and then it was time to go.

Outside, Kenny looked over and saw Burgio talking to someone from the neighborhood. Kenny said nothing and headed up the street. He and his friends were nearing the car when he suddenly turned around. "Something came over me that I wanted to go over and confront him." It was the toxic blend of his thoughts, the drinking, and "all the frustration built up." He'd snapped.

Burgio saw him coming. "Kenny, how you doing?"

"How am I doing?" Kenny asked rhetorically. "I'm not doing too well, Jimmy. I'm going to jail for thirty-four months because you're a fucking coward."

"Is this where you want this to go?" Burgio said.

Kenny did. "You pussy," he yelled. "You should get up and speak like a man and stop hiding behind things."

"You don't know what you're talking about." Burgio's voice was expressionless.

"I don't know what I'm fucking talking about?"

Kenny was drunk and yelling. Burgio was sober and calculating.

"Go home before I P-C you." Burgio warned. He knew that

threatening to place Kenny in protective custody would stoke him further.

Kenny yelled wildly at Burgio, readying for a fight. By now his friends were surrounding him and pulling him away.

"You wouldn't want to do this when I'm in uniform," Burgio taunted.

Kenny said any time, any place.

"I'm off in twenty minutes. Pick a spot."

Kenny's friends hauled him away to the car, with Kenny yelling. "I'm going to jail for you, you piece of shit."

Burgio watched as the car drove off. He considered Kenny Conley all talk. When Burgio finished work at the bar, he saw no sign of Kenny or his friends. "Nothing happened." Smugly, he added, "He wouldn't want to fight me."

Kenny's friends had taken him home. By the next morning, Kenny was disgusted with himself for trading insults like a schoolyard thug. The one thing—Burgio never once denied the accusations about Cox. He never said, "I didn't do it." Instead, it was cryptic macho-posturing: "You don't know what you're talking about." Words that meant nothing.

In a weird way, Kenny was truly looking forward to Cox's trial; he was going to testify—finally tell his story. That was going to be a huge relief.

Maybe, then, Burgio would finally get his due.

Mike Cox was doing his best to make that happen, working with his attorneys to gear up for the civil rights trial. His lawyer Steve Roach had early on asked another Boston lawyer with extensive trial experience to join the case. Robert S. Sinsheimer was, in many respects, Roach's mirror image: intense—to the point of seeming hyperactive—indefatigable, and physically unimposing. Slight in build, Sinsheimer topped out at five-five, and Roach wasn't much taller. The Brooklyn-born lawyer had grown up north of Boston and then attended Dartmouth College, majoring in government and graduating in 1975. He then went to law school, attending Suffolk University Law School in Boston, where during his last year he was on the winning moot court team. Upon graduation in June 1979, he taught legal writing at Suffolk for two years, which was when he met Steve Roach, a student in his legal writing class. Sinsheimer then worked as an assistant district attorney in Plymouth County, gaining his sea legs in the courtroom, and beginning in 1984, he started out in private practice doing a mix of civil and criminal defense work.

Sinsheimer had far more trial experience than Roach, but that wasn't his sole appeal; he'd had a taste of what it was like to take on the Boston Police Department. In the early 1990s, he'd represented a man falsely accused in the murder of a drug dealer in Dorchester. During the trial, Sinsheimer shredded the credibility of the police investigation, and the jury quickly acquitted his client, deliberating for less than two hours. Sinsheimer afterward filed a civil rights lawsuit against the department. While accepting Roach's offer to help out in the Cox case, Sinsheimer was busy representing another man wrongfully convicted of attempted murder. He uncovered that police perjury—or testilying—had helped convict his client. By late 1997 he succeeded in getting the conviction thrown out. In his ruling, the judge condemned the police investigation, calling sworn testimony by officers "a fraud upon the court" and a "disgraceful episode."

The Cox case, then, was "right up my alley," Sinsheimer said later.

Trial preparations were growing increasingly intense during 1998, a hectic pace of analyzing police records, deconstructing the failed internal police investigations, and taking depositions from up to twenty police officers and officials. Sinsheimer and Roach sat through most of Ted Merritt's successful prosecution of Kenny Conley in June, looking for pointers on a plotline for their own civil rights case. They came away with an unexpected bonus when Smut Brown testified that Dave Williams hit Mike at the fence. "Brown gave up Dave Williams," Sinsheimer said, "and actually hearing Brown say Williams hit Mike was new." It was a eureka moment of sorts, and the lawyers knew they were going to call Brown to the stand in the civil rights case to do a replay of his testimony from the Conley trial.

Sinsheimer thought another eureka moment came when a top police official filed a sworn affidavit openly acknowledging the department's blue wall of silence. "Officers are reluctant to break the 'code of silence' and to testify against their colleagues," Ann Marie Doherty, chief of the Bureau of Internal Investigations, had written as part of her explanation for why departmental probes into the beating had failed. During three days of deposition in June 1998, Sinsheimer could tell Doherty wished she could take back the affidavit that helped Mike's effort to expose a police culture of lying. "It was an admission that would cost the department," he said. By then, too, Doherty was gone, transferred by Police Commissioner Evans to a new post overseeing the police academy. To succeed her, Evans chose Jim Hussey, who'd handled the Internal Affairs inquiry into the beating.

The taking of depositions was exhausting. Tempers flared. Over the summer, Mike sat through six grueling days of often pointed questioning from attorneys representing the officers, the police department, and the city—and on the seventh day he'd had enough.

"The record should reflect," said Tom Drechsler, Burgio's attorney, "that Mr. Cox has just left the room without asking for a recess and he has gotten up and left in the middle of a question."

Steve Roach quickly came to his client's defense: What do you expect? "You're browbeating him," he said. "Mr. Cox was visibly upset and he left the room to take a break."

Drechsler and the others denied any such thing; countering, they accused Roach of using hand signals to coach Mike on how to respond to their questions. "It's prompting his client," Drechsler complained. "It's inappropriate. The record doesn't reflect the gestures."

Roach wouldn't give an inch. "Questions in the now seventh day of this deposition have been abusive," he charged. "They have been repetitive."

The gloves were off. Each side got nasty. "Oh, please," Drechsler said, fed up with what he considered Roach's persistent interference. He called Roach's conduct "highly unprofessional and highly inappropriate."

"You have a very suspicious and paranoid mind, Tom," Roach said.

"Well, excuse me, but I don't need personal insults and criticisms from you."

"Well," said Roach, "that's what you're doing to me."

Drechsler admonished Roach. "Don't use words like 'paranoid' and things like that unless you're a qualified psychologist and you're prepared to testify on the record. That's a personal insult. I'm not getting personal with you and I'd appreciate and expect for you to refrain from personal insults, okay?"

But the fireworks did not let up. For the remainder of Mike's final day of deposition, Roach stood guard, constantly interrupting and challenging his opponents in a bid to block and parry a beating by words from the phalanx of attorneys.

The opposing lawyers were not impressed with Roach's lawyering. They seemed to consider Roach out of his league. "We'll take your lessons on trial practice another day," one smartly told him.

Indeed they would—come December with the start of Mike's case.

Mike Cox had never thought much of the diagnosis made soon after his beating that he suffered from post-traumatic stress disorder (PTSD). He didn't much believe in psychiatry, and, to him, the best therapy would be winning his federal civil rights case. Even so, his lawyer Steve Roach took him to a second psychiatrist for another opinion—both for Mike's own well-being and for the purposes of the lawsuit. Roach personally drove Mike twice to the doctor's office in the bedroom community of Chelmsford, thirty miles north of Boston. Mike and the Harvard-trained forensic psychiatrist spoke privately while Roach waited outside. The doctor also met with Mike and Kimberly together.

The psychiatrist found Mike friendly and cooperative, but a bit guarded. He assessed Mike's intelligence in the "superior range," with no indication of delusions, hallucinations, obsessions, or compulsions. "His insight is limited, however," wrote Dr. Ronald P. Winfield in a report prepared in May 1997, "in that he continued to feel and to state that he did not believe that he had a psychiatric disorder."

Mike's views notwithstanding, Winfield affirmed the earlier diagnosis for PTSD, a disorder he determined was "directly due to the beating which he suffered at the hands of fellow Boston Police Officers." Observing that Mike's symptoms had persisted in the several years since the beating, Winfield said his condition was "chronic."

Of particular interest, the psychiatrist found that Mike's "pre-existing psychological makeup made him particularly vulnerable." He cited a study showing that Vietnam War veterans who'd been gung-ho about the war prior to combat were more susceptible to developing PTSD. "The destruction of one's beliefs is an intensely painful emotional blow," the psychiatrist noted.

Winfield then compared that dynamic to Mike's experience. Before he was beaten, "Michael Cox believed in the American Dream; he believed in the goal of a color-blind society. Michael's personal and professional demeanor reflected these values, and he incorporated them into his life: He sought success through education, perseverance, and hard work."

The pummeling at the fence—and his abandonment by fellow police officers—had changed all that. "To him, it was a violation and a repudiation of the Dream by which he had directed his life.

"Like Vietnam Vets, Michael was exposed to a trauma that at once endangered his life and undermined his beliefs."

The psychiatrist also minced no words in emphasizing the singular power race played in his assault. Trauma, explained Winfield, was defined clinically as the experience of being made into an object. While a beating of this sort would be difficult for anyone to process, "objectification has a special depth of meaning, and emotional resonance for black Americans." During two centuries of slavery, the psychiatrist noted, blacks were considered objects, "chattel; i.e., an item of property."

For Mike, concluded Winfield, the trauma as a black man "being made into an object" was therefore "especially intense, destructive and psychologically malignant: psychological ghosts of night riders, lynchings and Jim Crow were resurrected."

During the evening after Kenny Conley's conviction in June, the crank calls resumed at the Cox household. "You're an asshole," a voice told Mike. Mike hung up. Later in the night the phone rang again, but Mike hung up as soon he recognized the gravelly voice. It made him furious that even though he was the one who'd been beaten, somehow, in the perverted logic of the cop culture, he still was the wrongdoer for pushing for justice. Now a cop had been convicted of perjury and it was supposed to be his fault.

He didn't know Kenny Conley, but he resented the support Conley was getting, whether nefariously—as with the crank calls—or above board: the fund-raisers, the "Time for Kenny," and the media interest in his plight. No one had staged a rally or fund-raiser for him.

"The support I received has been quiet support by good friends and family, you know, a few people within the department," Mike said. "His support has been overwhelming."

In truth, some of the reasons for the disparity were less about the men and more about their neighborhoods. Roxbury was a neighborhood splintered socially and politically, while Southie's cohesiveness was legend. Roxbury had nothing matching "Southie pride." Early on, too, Mike had rejected the few overtures made by some of Roxbury's political and religious leaders; he was private by nature and wanted no part of turning the incident into a political or racial cause. Kenny Conley, meanwhile, with his friends' help, had gone public unabashedly with his insistence of innocence.

Importantly, part of Kenny's public crusade was to make clear he was no cover boy for the police cover-up. He railed against the assault and the mistreatment of Mike Cox—and that included harassing telephone calls, tire slashing, or any form of ostracizing Mike. Kenny couldn't control anonymous cops who mistakenly thought he was standing tall for the blue wall. What he could do was openly criticize the inability to get to the bottom of the beating. "I'd like to get across to Michael Cox," he told reporters, "that I had nothing to do with it and if I could have helped him I would have."

Cox and Conley: They'd become an odd couple. One from Roxbury, the other from Southie. Prior to January 25, 1995, they were young cops with nothing but bright futures ahead. Then they "met" at the fence on Woodruff Way chasing Smut Brown. The utter failure to solve Mike's beating had derailed both their lives: Mike was the pariah—he'd protested too much; Kenny faced thirty-four months in prison due to a misguided federal prosecution. Both outcomes resulted because the beaters and eyewitnesses went mute. Cox and Conley were now unlikely allies against the blue wall of silence.

It was unbelievable to Mike, the glares he'd gotten from the off-duty cops in the courtroom while he testified at Kenny Conley's trial, the cold shoulders he got when walking the halls of police headquarters. When he supervised a detail of officers working a Red Sox game at Fenway Park, he got looks from some of the cops under his command. He'd overhear bits of their muffled talk. "He's the one," or, another time, "That's the one, Cox." Mike used to like doing the games; he'd work an average of about four a month. But no more. Even though the Sox, led by their ace, Pedro Martinez, opened the 1998 season on fire, going 18–8 in April, Mike quit signing up for the Fenway Park details. They made him too uncomfortable.

His wife and family wanted him to quit. One sister-in-law living in Michigan who worked at the Ford Motor Company mentioned she could help him find work with the automobile giant; a sister who worked at Gillette headquarters in South Boston said the company often hired former police officers in corporate security; the father of one of his son's teammates on a youth baseball team was a senior executive at Fidelity, and he offered to help land him a position with the financial services company. In police circles, he was approached about joining the Massachusetts State Police or the FBI. Each time

Mike said, No, thank you. His dream about police work in Boston may have been shattered, but he wasn't going to walk off the job. First, he needed the paycheck—he had three kids and a wife just starting a medical career—and while he couldn't quite explain it or say it made sense, he felt "being a police officer I'm also a lot safer."

So Mike stayed on, although his work performance was on a steady decline. He struggled to keep up with the cases assigned to him in the Internal Affairs Division, and he was regularly apologizing, he said, "for lagging behind." It wasn't just the stress and distraction of the looming civil rights case; persistent headaches dogged him, and he had trouble concentrating. It seemed to take forever to write up his reports. Looking for relief, he was taking daily doses of Duradrin, a powerful prescription painkiller for migraines. By the summer, Jim Hussey, his boss, decided to lighten his workload.

When one of his mentors tried to recruit him to apply for an opening in the homicide unit—to work on a new squad looking to solve cold cases involving the street gangs—Mike said no way. The old Mike would have jumped at the choice assignment, but he'd decided he couldn't work again on the frontlines. "It has to do with my ability to trust the individuals that I work with," he said. "Trusting people with my life." Too much had happened, and he avoided police officers and police talk. "I'm not the person you would want to hang out and be around if you want to improve your career."

He was looking to lower his profile, not raise it. Instead of the prestigious homicide unit, he expressed interest in an opening in an obscure, pencil-pushing unit within the IA division—the audits and review section. "They go around and do audits of the drug unit, the seized money section, the tow log," Mike said. It was the antithesis of the old Mike to prefer paper over street action. The audit section was buried inside a division that was already isolated from mainstream policing. But that was its very appeal. "It involved, I'd say, less contact with police officers and their supervisors, less people," Mike said. "More isolated." Mike was trying his best to be the invisible man.

Not all the calls to the house after the Conley conviction were hateful. Jim Hussey checked in to see how Mike was holding up. He wasn't surprised by Mike's reticence. "He's not a man of many words," Hussey said later. The two chatted, and Hussey said he wanted to resume surveillance of Mike's house for a few days. Hussey was well aware of the crank calls and other harassment, and he wanted to take "all precautions necessary, just in case there was a loose cannon out there." Hussey was hoping to make Mike and his family feel more comfortable. What did he think?

"That will help me sleep better." The line was more an attempt at dry humor than the truth. Mike never slept well. He stayed awake worrying about his family's safety.

In the more than three years since his beating, he still had not explained to his sons what happened at the fence. Police officers are "good, they're your friends," he'd always told his boys. "How do you explain to a child that you were beaten and basically left for dead by other police officers? And then other police officers witnessed it and no one has said anything.

"How can you tell a child that?"

No one knew better than Kimberly the continued toll on Mike.

"Before," she said, "he was this nice, easygoing person who enjoyed doing things with his family. You know, he would joke a lot, he had fun. We did things together.

"I'm referring to the person that he once was. I'm referring to the fun-loving person who wasn't paranoid, wasn't depressed, wasn't irritable, wasn't difficult to get along with."

Kimberly's comments came during her own deposition. The session began at 9:23 A.M. on August 6, a sunny morning with temperatures in the mid-70s, and ended six hours later. The questioning was conducted in the plush downtown offices of the private law firm hired by the city and the police department.

It marked the first time Kimberly had spoken extensively about Mike, and before the first question was asked, Steve Roach sought to assert the ground rules. Given the "abusively long deposition with her husband," Kimberly was "only going to be here one day. We feel one day is enough."

"Let's see how we do," replied the city's attorney.

Kimberly began by summarizing her upbringing in New Orleans, her meeting Mike in Atlanta, their marriage, family, and respective careers in Boston. Most of the six hours then became a chronicle of a troubled Mike Cox after January 25, 1995.

He was different now, she said, and seemed depressed: "Lack of appetite, always feeling exhausted, wanting to sleep, not wanting to have company over. Wanting to be alone. Just not participating in daily family life like he used to."

He used to coach his boys in sports, but had stopped. He used to read to them at bedtime, but now rarely did. "He was much more physical with them before," she said. "They're boys, they like to be tough, they like to wrestle, and he doesn't do that anymore."

In January they'd celebrated the birth of their first daughter, Mikaela, but, she said, "When he walks in he takes her and hugs her and kisses her, and then usually he gives her back."

It was all such a sharp contrast to the early days of their marriage, when she was commuting to Philadelphia to attend medical school and they'd not only successfully met the challenges of work and home life, but had grown closer.

Now Mike seemed only partly there. "If we're having a conversation he'll walk out of the room in the middle of the conversation. I'm talking about one thing and he'll leave that subject and go to something else, or he'll pick up the phone and he'll, you know, start dialing, calling someone on the phone and, like, Hey, we're talking."

He could be quick-tempered and unpredictable. "I hate to seem trivial, but just last night I went to the mall to buy some stockings, and that was a big deal.

"He said, Why did you need to do that? Well, I had to go get some stockings, and he had to have a fifteen-minute, you know, argument over the stockings. To me, that's crazy.

"Every little single thing, every single day." Mike was always turning off the kitchen ceiling fan with its four bright lights. "You know, we argue every day about sitting in a semi-dark house because the lights hurt his eyes."

Like Mike, she worried about their safety, but she thought Mike had become obsessed and hypervigilant. "Worrying about what time I get to work, what time I come home from work, what time the kids get home.

"He's preoccupied with making sure the doors are locked, rechecking them at night."

She certainly had her opinion about the source of her husband's struggles. "He feels abandoned, and basically there's been no real way for him to feel that he's received justice about what happened to him.

"I feel that in his mind, if someone had been identified, if there was someone who would take responsibility, you know, for their actions for what was done to him, he could at least have some of—well, there's closure to this."

The lawyer finally asked, "Do you and Michael have any plans to separate."

Kimberly had hung in there, answering the relentless questioning seeking intimate, painful details about the more than three years since Mike's beating. It was like picking at a wound over and over; she'd admitted the marriage had become tense, strained, and, at times, seemed "unbearable." But it was as if this question by the lawyer had gone too far, and her back went up.

Firmly, she replied, "No. We haven't made any plans to separate."

The lawyer and deponent then had their most curt exchange.

"You still love Michael?"

"Yes, I do."

"He loves you?"

"Yes."

The next month, during the early evening of September 14, Smut Brown had some urgent business to take care of, so he asked some pals to drive him over to Sutton Street in Mattapan. Sutton was a mostly barren one-way street about a block long. It was lined with worn triple-deckers, unkempt yards, and drug dealers. Smut knew full well about the drug activity; following his acquittal the year before in the Lyle Jackson murder trial, he'd resumed his own illicit activities, associated with the outfit known as KOZ.

But this was not about business; it was a family matter. His mother, Mattie, was staying in a third-floor unit at 5 Sutton Street. She was with one of his big sisters and her young daughter. The previous night, his mother had awakened in the middle of the night to partying and loud music in the unit below. She had gone down and complained. One man followed her back upstairs. "He was drunk," Mattie said later. "He came into the apartment cussin', rude and disrespecting me."

When Smut learned what had happened, he got angry. When he arrived at the house, he climbed the six cement steps and confronted the man on the front porch. His mother was not there; she'd gone out shopping in Mattapan Square. Smut saw the encounter as a teaching moment—about respect—but it wasn't long before the street talk turned into a street fight. Smut was getting the better of the fistfight when he noticed the guy was carrying. They began struggling over the weapon, and it fell to the ground. "Then this guy's friend picks up the gun," Smut said. The man was pointing the gun at Smut and his friends, and "I'm freakin' out, thinking, man, now I've gotten my friends shot when this was my beef." He pushed the man he was fighting into the gunman. Then Smut ran off the porch around to the side yard littered with trash. The gunfire began.

"I got a call on my cell phone," Mattie said, "that Robert got shot."

Smut was hit once in the back but kept running. Police and emergency medical workers found him lying in blood in the middle of Sutton Street at 7:19 P.M. He was taken to the emergency room at Boston City Hospital, where his mother joined him.

The shooting made the newspapers. "Robert Brown is the only person, other than Michael Cox himself, willing to come forward to testify about what happened that night and who can help us identify specific Boston police officers," Roach told a reporter from the *Boston Globe*. The attorney was openly worried about Smut's availability for the upcoming civil rights trial.

Smut was lucky, though. He quickly stabilized and, after five days, was released. "The only time I wasn't there," his mother said, "was when Indira was with him. I went home and changed clothes and came back. I never left him alone." Smut left the hospital with a souvenir—the bullet in his back. Doctors decided not to remove it. He left with something else too—a fondness for painkillers. He was soon a heavy user of Percocet.

The next month, Smut was shooting pool in a hall in Dorchester when he ran into Craig Jones and Mike Cox. The two were apparently checking up on Smut's recovery as part of their own trial preparation. Craig, Smut said, "was like asking, 'Yo, you know, you gonna come to the civil trial 'cause he, you know, he needs you.'"

From Smut's point of view, he'd gotten nothing but trouble for testifying. The Conley trial had left him confused, the way his testimony was twisted, but the big picture wasn't his concern; to the contrary, he wanted to get in and get out. Keep it as simple as possible. He didn't even want anyone knowing about his involvement. "I gotta live on the street," he said, "and I don't want people seeing me with the police and they thinking that I'm doing something that I have no business doing, and I end up getting found dead somewhere." If not for giving his word he would have told Craig and Mike he was done with Woodruff Way. But he'd told Bob Sheketoff he would do the right thing, and he was going to finish what he'd promised. Yeah, yeah, Smut said. "I'm gonna be there."

By the year of the trials—Kenny's trial for perjury and Mike's upcoming civil rights trial—the Boston media had finally ended its Big Sleep and was following the Cox scandal extensively. Columnists and editorial writers at the city's two major newspapers, the *Boston Globe* and the *Boston Herald*, joined the news coverage with increasingly pointed commentary about the failure of Police

Commissioner Evans and other law enforcement officials to break through the blue wall to bring the beaters to justice.

For his part, Evans demonstrated a keen interest in public opinion. Following a series of police screw-ups, ranging from the tragic death of the retired black minister during a mistaken drug raid to the uncovering by the *Globe* of years of corruption by two veteran officers, Walter "Mitty" Robinson and Kenny Acerra, he wrote an op-ed piece in the *Globe* heralding his new public integrity and anti-corruption initiatives. He said the inference in media coverage that his administration was not interested in rooting out wrongdoing was unfair. "This impression cannot be allowed to go unaddressed."

Not long afterward, he went on the public relations offensive to counter the *Globe*'s ongoing investigation into Boston police misconduct. For a meeting at the newspaper's headquarters with editors and reporters, Evans brought along data showing a statistical improvement in the department's self-policing. "One thing I feel strongly about is that we have to clean up our own problems," he told the roomful of journalists.

Then, in the wake of another *Globe* exposé in late 1997 about testilying, Evans joined court officials and Ralph Martin, the district attorney for Suffolk County, to announce a crackdown on police perjury that could result in botched trials, wrongful convictions, and few repercussions for officers caught lying. They unveiled a new reporting system by which judges who suspected an officer was lying would notify police and prosecutors for a follow-up inquiry. The plan seemed a step forward and made for a great sound bite. But the talk was more stunt than substance. "I honestly don't remember anything about this," Ralph Martin said years after the much-ballyhooed reform was first announced. It turned out the reporting system was never implemented.

By the fall of 1998, the *Globe*'s editorial writer Larry Harmon, who specialized in criminal justice matters, was leading the growing view that Evans and all the other law enforcement agencies had mucked up the Cox case—a tragic comedy of errors and lack of will that hit a new low with the criminal conviction of Kenny Conley. "The US attorney is supposed to squeeze witnesses and send tough messages about the fate of any officer who commits perjury or tries to cover up knowledge of a crime. But he is supposed to squeeze the right people," read the *Globe* editorial of October 10, 1998.

Then came the dart aimed at the police commissioner. "While Conley readies for prison, at least three Boston police officers who likely have greater knowledge of the case—James Burgio, David Williams and Ian Daley—remain on the job."

The *Herald* columnist Peter Gelzinis was also unimpressed with the department's track record regarding Cox and Conley, saying Cox had become an outcast in "a department that's never been able to locate its moral backbone with this case.

"It is Michael Cox who gets shunned, not the officers who beat him. Not the officers who are directly responsible for stripping Kenny Conley of his career at 29."

By the end of the month—and a few weeks before the start of Mike's trial—Evans finally took action that most familiar with the scandal considered long overdue. Burgio, Williams, and Daley were stripped of their badges and weapons and placed on administrative leave. Evans included a fourth, Kenny's partner, Bobby Dwan, a be-wildering move that left Bobby stunned and apoplectic; in short order, he hired one the city's best lawyers to fight back.

Evans's announcement made front-page news. But the commissioner was noticeably circumspect when asked why now? Why did he wait nearly four years after the beating to put the officers on paid administrative leave now?

"It's a combination of a lot of things, a lot of information, all coming together, pieces from different investigations," he told reporters. The implication was that new information combined with a dogged determination had created a critical mass requiring action. But no amount of spin could quell the notion the announcement was window dressing, a public relations move just before a trial where the department itself had been accused of engaging in a "pattern of indifference" to excessive force and the cover-ups that followed. In fact, in terms of the actual evidence, nothing of substance had changed or improved since Bob Peabody's probe had ground to a halt two years earlier. The truth was Evans could have taken the officers off the street a long time ago.

Mike Cox certainly saw the action that way—as a stunt. "I was pretty infuriated," he said. "It seems hypocritical to do this on the eve of a civil trial. It's another move on their behalf to spin the best light they could." His attorneys were angered as well, but Rob Sinsheimer saw Evans's announcement as a "crack in their armor." The timing, he said, looked so bad. "It made them look like they were panicking, and it emboldened us."

Settlement discussions are part of any lawsuit, and during Thanksgiving week the pace of the talks picked up. Sessions were held in the jury room of Courtroom 18, where the trial was scheduled to start in early December. The presiding judge, William G. Young, attended, urging the parties to find a common ground. He sat at the conference table across from Mike Cox, and, while lawyers did most of the talking, the veteran jurist couldn't help but notice Mike's "quiet dignity." Sinsheimer was always agitating the city's attorneys with his nickname for the lawsuit: "Boston's Rodney King case." From the judge's perspective, the city was offering relatively substantial money—a figure the newspapers put at $300,000. But there was a sense in the room that for Mike Cox it was the principle that was at stake, not financial compensation. The talks never came close to reaching a settlement—an impasse that seemed further proof of Mike's feeling about justice and accountability.

The next day, the day before Thanksgiving, the judge issued a series of rulings setting the stage for the trial to finally begin. He cleaned up the case a bit, dismissing parts of Mike's lawsuit, such as the claim that Police Commissioner Evans was personally responsible for violating Mike's rights, as well as the charge the city had failed to provide him with medical care. But he rejected the police department's bid to have the entire case against it thrown out. "The motion of the city for summary judgment is denied," he said. "It's denied as to their liability for failure to train, for failure to supervise, and failure to discipline. All of those matters warrant a jury trial and a jury trial will take place." He also ruled Mike's civil rights were, without question, violated by officers who beat and abandoned him. "I have a number of Boston police officers beating a suspect, then abandoning him in the dirt after discovering he was a police officer." The fundamental question facing the jury, then, was whether the officers, the police department, and the city were liable for those violations of Mike's civil rights.

Citing reasons of fairness, the judge announced he was going to split the case into several jury trials: the first would be against the four officers, the second against the city, and a third, if necessary, to assess damages. "We cannot try the individual defendants along with the city," the judge said. "We cannot do it because evidence properly admitted against the city is so prejudicial to the individual defendants."

The judge then further refined the first trial in response to Willie Davis's concern that evidence against the three accused in the beating would spill over in a detrimental way against Kenny. The solution, the judge said, was to have Mike's lawyers first introduce evidence against Burgio, Williams, and Daley and then offer evidence against Kenny. Davis and his cocounsel, Fran Robinson, would have preferred a separate trial, but the judge's ruling didn't seem so bad. It meant the judge would frequently be instructing the jury that a witness's testimony was not evidence against Kenny but against the other three—a distinction, said Fran Robinson, that might come across as a signal from the judge that Kenny was not one of the bad guys.

The session ended with some housekeeping. "I sit nine to one," the judge told the lawyers. "There's a reason for that." He didn't want to waste the jury's time with drawn-out legal haggling about the admissibility of evidence or other legal concerns that inevitably arise during a trial. "We'll talk about those things in the afternoon," he explained. Moreover, the afternoons would provide the lawyers time to catch their breath.

With that, the civil rights action, now regarded as one of the biggest in the history of the Boston Police Department, was ready: Jury selection was slated for December 7. Mike's lawyers were at once confident and anxious. Sinsheimer, for his part, was eager to finally get going, believing opposing counsel for the cops and the city had underestimated them. Up against a Goliath, he thought he and Roach "made a great team. Steve was the best thing that ever happened to Mike Cox because of his attention to detail."

What had begun one night at a fence on a dead-end street had grown into a major crisis for the police department, its commissioner, the city, and its mayor. The case had forced Boston to examine its police culture and racism. Would there be justice for Mike Cox? For Kenny Conley? For a city that, with its nasty racial legacy, was heading into a new century? Mike Cox was certainly tired of waiting. He'd realized long ago he could not depend on the police department for the truth. He was on his own, and he was ready. The trial, he said, "is the only forum I have to try to get the truth out."

Mike was hoping for the best, and by that he didn't mean hefty monetary damages. "I hope this case will change the department," he said. "Because if it doesn't, I would have lived through a terrible ordeal and no one will gain anything from it."

CHAPTER 18

The Trial

Steve Roach, dressed in a dark suit, stood in the well of the courtroom. Even though he was the least experienced, he and cocounsel Rob Sinsheimer had agreed he would start them off. The Cox case was his baby; he'd lived with it for nearly four years and, in the days leading up to the December 7 start, had rehearsed hard for this moment.

He moved deliberately across the thick blue carpet and faced the twelve jurors.

"Good morning," he said.

"Good morning," the jury of eight women and four men replied in unison.

Roach paused. He looked at them. Then he began: "Ladies and gentleman, this case is about police brutality, police cover-up and a blue code of silence."

The jurors, seated in two rows in wooden chairs, listened intently to every word.

The Cox trial was among the first to be held in the new federal district court built on the waterfront in South Boston. Overlooking Boston Harbor, the facility featured a brick dome and curving glass wall that, one architecture critic noted, "falls freely as a sail for seven stories, with nothing touching it, rigged with turnbuckle stiffeners like those on a ship." The corridors on each floor were open balconies bathed in sunlight. "They line the glass wall like boxes at the opera," the critic said, "looking out to a magnificent drama of harbor and skyline."

On the fifth floor, Courtroom 18 was the domain of Judge William G. Young. Young was a history buff who saw that the fixtures in his oak-paneled courtroom replicated those in one of the oldest courthouses in the state—or nation, for that matter—in Newburyport, Massachusetts, a seaside town north of Boston. When he was a judge in the state Superior Court, the Newburyport courthouse had been his favorite. Young applied some personal finishing touches as well. Instead of the portraits of jurists that typically adorned courtrooms, he hung two oil paintings by his father. Both were of ships at sea—*Spinnaker off Target Rock* and *Whale Ship Emerald.*

The Long Island, New York, native was a graduate of both Harvard College and Harvard Law School. During his seven years as a state judge, he presided at the "Big Dan's" trial in which four men were convicted of raping a woman in a bar. Jodie Foster later starred in the movie about the case. In 1984, President Ronald Reagan appointed him to the federal bench.

Like the jury, the judge settled in to listen to Steve Roach. Young wore wire-rimmed glasses and shifted comfortably in his inky robe. The elevated bench hid the fact that he was short, maybe five-five, which came as a bit of surprise for someone whose voice carried so clearly and sonorously. Looking out, he saw a courtroom filled to capacity. Five oak tables in the large well were all in use. Roach, Sinsheimer, and Roach's associate had one. The various defense attorneys occupied the others—Kenny Conley's lawyers Willie Davis and Fran Robinson; Ian Daley's attorneys; Dave Williams's attorney; and Jimmy Burgio's attorney. None of the other defense attorneys wanted to be at the same table with Burgio's attorney, Tom Drechsler. The judge had ended the fuss by assigning seats.

In the gallery beyond the oak railing, Kenny sat off to one side, making sure to stay away from Dave Williams and the others. He wanted nothing to do with them. Williams sat to the right, his brow furrowed. Ian Daley was seated in another pew. The one face missing in the crowd was Jimmy Burgio's. He was nowhere to be found. In civil cases, defendants are not required to attend, and Burgio stayed away. Mike and Kimberly Cox sat in the front row directly behind their lawyers. Family, friends, off-duty cops, and reporters filled the oak benches set up in five rows.

Before making way for Roach, the judge took a few minutes to talk to the jury. Chosen first thing in the morning, the jury included a bank teller, a bartender, a teacher, two retirees, and the manager of a restaurant. One, Carol Goslant, was an electrical engineer; Bob McDonough worked for Titleist, the golf ball manufacturer, in the training and development department; Sharon Schwartz was a homemaker.

"You twelve men and women are the judges of the facts," the judge said. He explained Mike Cox had accused the defendants of his beating and that it was his burden to prove their liability by a "fair preponderance of the evidence." In other words, his evidence implicating the accused must be "more likely to be true than not true." The judge made clear to distinguish Kenny Conley. "He's not suing Mr. Conley on the excessive force theory." The case against Kenny involved two claims—that he saw the beating and did not intervene to break it up and that he participated in a cover-up afterward. "The law forbids police officers from engaging in a cover-up," he said. "That means that we are entitled to expect police officers to play it straight."

The judge told jurors they could take notes. They could ask questions by passing a message to him through his clerk. Finishing up, he explained the lawyers would next deliver their opening arguments—"the guide book of what evidence they hope that they can put before you."

Judge Young was outspoken in his criticism of trial lawyers who styled themselves as courtroom thespians. He hated that some studied acting. "I'm actually quite hostile to the idea of acting lessons," he once said. Lawyers, he said, were principally teachers—"teachers of the facts." It was a demanding task, requiring skills that were "much less than those of an actor and much more the skills of a dedicated teacher. Not glib, not histrionic."

The judge didn't have to worry about Steve Roach. For the next fifty minutes, Roach told jurors in workmanlike fashion about Mike's beating and why the defendants were culpable. "We're going to show you that these four defendants violated Michael's civil rights as a result of a brutal beating that took place in the early morning hours of January 25, 1995 and that there was cover-up afterwards."

Using a map of Woodruff Way blown up into a poster, he pointed out the route of the police chase snaking through the city and then the lineup of cars screeching to a halt at the dead end. He said Williams hit Mike at the fence from behind with either a flashlight or a baton, Burgio kicked him in the face, and Daley got in some licks as well. During the "savage onslaught," once they realized Mike was a cop, "They ran. They took off. They abandoned him." That's the moment the cover-up began.

"They knew they couldn't claim that this was a suspect who was resisting arrest." Conley joined the cover-up, he said, by denying he'd seen Mike at the fence—a denial that was the basis for his recent perjury conviction in another trial.

Roach's own nervousness seemed to show when he faltered while describing Mike's injuries. "He still has traces of urine in his blood and now it's four years later." Realizing the error, he started over. "He still has traces of blood in his urine, rather."

He never mentioned the blue wall of silence by name, but Roach made clear that police officers were of no help when it came to determining responsibility for the assault. "Not one cop who streamed in that night will come here and say they saw Michael Cox being beaten by a police officer," he said. For emphasis, he added, "Not one."

Their case, then, was largely circumstantial. The strategy he and Sinsheimer had devised beforehand was to circle in on the accused by a process of elimination—showing that the assailants had to be the cops they'd accused. To do that, Roach and Sinsheimer decided to rely on Joe Teahan and Gary Ryan, Mike's gang unit colleagues. Teahan's account that he and Ryan were in the fourth or fifth car to arrive at the dead end meant the beaters were in the few cars ahead of them. Roach wanted little to do with the fact their accounts were at times contradictory, understanding that he could best seal the deal at the dead end by embracing Teahan and Ryan's basic storyline about finding Mike alone and bloodied—the functional equivalent of the yellow police tape used at crime scenes.

"One key fact that I would like to point out," Roach said, "is that Officers Ryan and Teahan closed the universe." He meant that the only people who could have done the beating were the cops who were at the dead end at the time of Mike's assault: Craig Jones; Richie Walker, who chased Smut Brown; and then Jimmy Burgio, Dave Williams, and Ian Daley. "The only people who were there who could have done it," Roach said, "are the three defendants, Daley, Burgio and Williams."

The point was Roach's principal and final one: "Ryan and Teahan closed the universe on the people who were there," he repeated. The evidence in their case, he concluded, will show "that it had to be at least those officers who beat him and at least those officers who covered it up. Thank you."

Sinsheimer welcomed Roach back to their table. He thought Roach had been terrific. "The culmination of years of work, succinct, boiled down. There wasn't a phrase in there we weren't ready to prove." Roach's words were also the last for the opening day of the trial. It was nearing one o'clock, the hour Judge Young had marked to end.

"Keep your minds suspended," the judge advised, a reminder there was still plenty to come. Even so, Sinsheimer was pleased the jurors were heading home with Steve Roach's words percolating in their minds as the first impression of the case to come.

Everyone in Courtroom 18 was well aware the case was unfolding along with the holiday season, and Judge Young made clear to all of the attorneys his expectation that the trial conclude before Christmas. If ever there was a judge who could deliver on that promise, it was Young—a stickler for preparation and efficiency. He regularly discussed with the attorneys the progress of the case and the time needed for witnesses.

The next morning began with a quick succession of opening arguments from the attorneys for Williams, Daley, and Burgio. They went about attacking the circumstantial case Roach was pledging to present against their clients. It wasn't close to being as clean and clear as Roach portrayed, they argued, and was bereft of direct evidence, except for the testimony of a drug dealer named Robert "Smut" Brown who had zero credibility.

The first up was Williams's attorney, Dan O'Connell. The veteran trial practitioner had replaced Carol Ball as Williams's lawyer when Ball became a state judge. "It's my privilege to represent Dave Williams," he told the jury, "who's seated over there, a Boston policeman." Each defense attorney seemed to stake out certain themes, and O'Connell's was that the plaintiff's evidence was a jumbled mess. "Mr. Roach wants, has indicated that this evidence is going to show certain things happened," he said. Except there's a problem: "It's internally inconsistent depending on which witness talks."

Next up, Ian Daley's attorney, Tom Hoopes, went after Mike. He said Mike and his lawyers had come to court acting as if "they wear the white hats." Not even close, he said. "Things are not as white or black as the plaintiffs suggest." He called Mike Cox and Craig Jones "two cowboys" for cutting in front of Daley to become the lead police car. They violated a rule that unmarked cruisers not lead chases and put themselves in "harm's way," he said, implying Mike Cox was asking for trouble.

Like O'Connell, Hoopes called the case wrongheaded, forcing his client to now defend himself despite his innocence. "Why? Because Mike Cox has brought the suit that you sit on here today. Seeking what? Seeking money." Mike, he said, was money grubbing. "I'm going to come back at the end of this case, ladies and gentlemen, and ask you one thing. I'm going to ask you not to let gray become green. I'm going to ask you not to let money color, subvert, and twist justice and the truth."

The gloves had come off. Tom Drechsler, representing Jimmy Burgio in absentia, jumped right in where Hoopes left off. "We don't want to be here. We didn't bring this lawsuit." Mike Cox, he said, "has brought this lawsuit for the purpose of getting money, financial remuneration." Drechsler ridiculed Roach's bid to use Teahan and Ryan to seal the dead end. "This is not a closed container." Police continued streaming down Woodruff Way and were running up from below on foot through a hole in the fence. "Municipal police, state police, Milton police, MBTA police, Housing Authority police, security guards—all were present at this location this particular night."

Drechsler said the evidence Roach was touting was smoke and mirrors—completely unreliable, given the wild and fast-breaking events at the dead end—"a very chaotic, confusing, dark situation." Sure, Mike Cox was beaten, but those guilty of the assault fled like ghosts into the night. Jimmy Burgio was a hero, not a beater. "My client did nothing more that night but make an arrest of one of these desperadoes, these murderers." Most important, he told jurors, they would not hear a single witness linking Burgio to the assault.

"You will hear not one witness say they saw James Burgio raise his hands to anyone or strike anyone or hit anyone, no less Officer Cox. The only thing you'll hear evidence about James Burgio doing," he said, "is handcuffing a prisoner."

Forget what Roach told you. "This isn't a process of elimination."

Taken together, the defense was a combination of attack on Mike's motives and a chaos theory for the beating that would make impossible the jury's task of assigning guilt. It was certainly true Mike's case was circumstantial, and it was certainly true, as Drechsler noted, no witness would testify Burgio kicked Mike in the face.

This worried Sinsheimer. He could see why Burgio might be optimistic about his chances. Ted Merritt, despite naming Burgio a target in the criminal investigation, had never sought his indictment—a fact illustrating weaknesses in the evidence. Sinsheimer was worried about "getting over the rail" in their case. By that he meant getting past the possibility of the judge issuing a directed verdict for Burgio. That was when a judge threw out a case because the evidence was so lacking it was unworthy to even send to the jury for its consideration. With Dave Williams, they at least had Smut Brown saying he saw Williams hit Mike. With Burgio, the case was entirely circumstantial.

There was no looking back at this point. Sinsheimer believed, of course, the case was winnable, but not easily. The game plan, as Roach had told jurors, was a process of elimination, and the immediate need now that opening arguments were completed was to give the jury a powerful dose of the harm done to Mike Cox.

They called as their first witness Donald Caisey of the gang unit. The idea was to call a cop whose testimony would be brief and set the scene at Woodruff Way. Caisey filled that role. He testified about arriving at the dead end after the beating and coming upon Mike, seeing all the blood, and thinking Mike had been shot. "My experience with the injuries that he had is very similar to someone being shot in the head. Someone shot in the facial area where blood comes out of every hole or area in the face."

The stage was now set for Mike. With Kimberly in the gallery watching, Mike strode to the witness stand and was sworn in. His voice was soft, and the reporters, knowing this was a key trial moment, shifted in their seats to hear.

"Mr. Cox," the judge said, "could you pull your chair in a little bit?" It would not be the last time the judge would have to instruct quiet Mike Cox to speak louder into the microphone.

"Would you state your name, please, for the jury?" Roach asked.

"My name is Michael Anthony Cox."

Moments before starting, the judge seemed to have rattled Steve Roach by refusing his request for a brief recess. It would have been a chance for Roach to collect his thoughts, but now he had to hurry and jump right in. The rushed start seemed to show during Roach's initial handful of biographical questions.

"Where do you live?" Roach asked Mike.

"I live in Dorchester, Massachusetts."

Seconds later, Roach asked again, "And do you reside in Boston?"

"Yes."

"Which part?"

"Dorchester."

Roach led Mike through a courtroom rendition of *This is Your Life*. Following his lawyer's lead, Mike described growing up, his family, his marriage, becoming a police officer, and working with Craig Jones in the gang unit. After a while the judge grew restless with all the background. "Let's get to this case, Mr. Roach."

Roach followed suit. "Directing your attention to January 25, 1995."

For the next hour Mike held the courtroom transfixed with a narrative that began with an urgent call that a fellow officer had been shot at Walaikum's. Mike explained how he and Craig became the lead car at an intersection when the Lexus raced downhill and almost ran into their cruiser. He described the street clothes he wore that night, and, when Roach approached him with the long black parka, he stood to examine it. The jacket was admitted into evidence as Exhibit 15. He remained standing and, using a pointer, identified on the aerial blow-up of Woodruff Way the location of the Lexus and police vehicles. He stood again a few minutes later to demonstrate to the jury how he'd run to the fence after a suspect, "and I reached up like this to grab both of his arms."

Mike then described the first blow, and how, when he turned to see who'd hit him, he was hit hard again. That's when Mike's voice cracked. Tears welled in his eyes, which he wiped away. He fought the emotions, not wanting to lose control completely.

"Where on your head?" Roach asked.

"In the front of my head."

"Can you point to the jury where you believe you were struck?"

Mike pointed to his forehead. "It's about a two-inch scar up there."

Mike was next on the ground on all fours. He looked up. "I could see there was a person standing in front of me who had on boots." The man was dressed in a dark-colored uniform, "and he was white. And as I started to look up, that person kicked me directly in the mouth." The beating just kept coming, and he never saw his assailants' faces. Then he heard someone yell, "Stop, stop, he's a cop," and the blows stopped.

Roach asked Mike if he recognized the voice yelling stop.

"It was Dave Williams."

The testimony had momentum—vivid, horrific, and with few objections from the defense lawyers to offset the flow. The judge was the one who broke the spell after Mike mentioned Williams. "Is this a good place to stop?" he asked Roach. "It's one o'clock."

But Roach was feeling sure-footed now. He didn't want the day to end just yet, not before getting to the gore. "Just a couple of more, your honor?"

"Go ahead."

"Did you notice whether you had any blood on you?" Roach asked.

"I just knew I was bleeding from all over."

"Can you describe where all over?"

"Bleeding from my nose, my mouth," Mike said. "I don't recall being able to see too well, you know, from my head I knew I was bleeding from all over."

The jury would be heading home, tasting blood.

"Your honor," Roach said, "if you wish, we can stop at this time."

Juror Sharon Schwartz, the suburban homemaker, was moved by Mike's testimony. "His injuries were grotesque," she said. Bob McDonough sat there thinking "he was very believable." The engineer Carol Goslant thought Mike was a "strong witness in that I believed him; he'd been through a terrible ordeal." The next day's *Boston Globe* story read: "Cox wiped away tears as he softly recounted for a federal jury how he went from officer to suspect to brutalized victim in a matter of minutes."

While nationally Americans were closely following calls for President Clinton's impeachment for his scandalous relations with White House intern Monica Lewinsky, and while Bostonians were trying to keep up with a number of pressing local matters, from worry that their football team, the New England Patriots, might move to Connecticut, to concern about the sorry performance of public school students on the new statewide standardized tests measuring expertise in English, math, and science, Judge Young's Courtroom 18 became a world unto itself focused on Mike's case for justice.

The skies outside were overcast when Mike returned Wednesday morning to finish sharing his memories of the attack. In particular, he recalled for the jurors the moment Ian Daley tried to arrest him at gunpoint and then, once Daley realized he was a cop, muttered, "Oh shit, Oh my God." Mike remembered later being in the hospital when Dave Williams blurted, "'I think,' you know, 'cops might have done this.'"

Guided by Roach, Mike explained how memories slowly returned as the weeks and months passed; for example, he recalled Dave Williams had yelled, "Stop, stop, he's a cop," after running into Williams at a funeral that summer. Roach finished by asking whether he saw Williams sitting in the courtroom. Mike replied he did.

"Could you point him out, please?" Roach said.

"Sure. He's the gentleman in the front row, the third aisle over."

Williams sat expressionless, his tall, muscular frame folded into the bench.

Roach was finished but Mike was not; he stayed put to be cross-examined by the lawyers for Williams, Burgio, and Daley. One by one they attacked his memory, saying it was unreliable due to his extensive injuries. They noted that from the beginning Mike had always said he chased not one but two suspects toward the fence—

a memory everyone knew was incorrect and which, they argued, made all of Mike's testimony doubtful. But Mike's response was disarming; he didn't claim to have perfect memory of Woodruff Way. "That's what I recall," he said. "Now, if I'm mistaken, I don't know." In a way, admitting he could be wrong about some things made him more believable, not less.

Trying to impeach Mike through his memory was their best hope. They asked Mike about a prior occasion when he was mistaken by other officers as a suspect. Hadn't Mike been grabbed by a cop and then another yelled, "Stop, he's a cop"? Wasn't Mike conflating that incident with the pummeling at Woodruff Way? Wasn't Mike mistaken about saying he heard Dave Williams yelling that command during his beating?

Tom Hoopes went last, and he sought aggressively to hit Mike hard on two points he'd raised during his opening. But Hoopes ran into a roadblock as soon as he suggested Mike was partly responsible for the night's tragic outcome once he and Craig took over the lead position in their unmarked cruiser. The judge quickly interrupted the lawyer's revival of the cowboy theme.

"This is an incredible reach," Young said. He accused Hoopes of "wandering around as to whether someone violated departmental policy." Even if the rule was broken, a victim of police brutality "still has his constitutional rights." It simply had no relevance in the case, the judge warned Hoopes.

Hoopes simply switched to the other point of attack—greed.

"Sir," he asked Mike, "would it be fair to say that you and your wife have a debt of about $200,000 from her medical school bills; is that right?"

Roach and Sinsheimer jumped to their feet to object.

They didn't have to. The judge tossed a pencil into the air. Everyone watched the writing instrument bounce across the bench. Young summoned the attorneys to his sidebar. Mike watched from the witness stand. He was exhausted but had weathered the effort to break him down.

When the huddle broke, the judge turned to face the jury. "We're trying a straightforward matter here, a matter worthy of trial," he said. "But this reference to greed and reference to other motivations for bringing a lawsuit is improper. And I have instructed counsel that it's improper." The lawyers had been slapped around, hardly the kind of last word any defense would wish for during a key cross-examination.

Mike's lawyers now went about trying to circle in on Williams, Burgio, and Daley. Joe Teahan and Gary Ryan testified about discovering Mike alone on the ground, battered and bleeding. No one else was around, and Teahan said only four or five Boston police cruisers were already there.

"No more than that?" Sinsheimer asked.

Drechsler objected immediately, and the judge, agreeing, told Sinsheimer, "Do not lead the witness." But Sinsheimer had made his point. Teahan's testimony planted firmly the idea that the dead end was the "closed container."

Then they called the only witness who provided testimony directly implicating any of the accused. "I seen a black officer hit him over the back of the head with the flashlight," Smut Brown said to open his account of the scrum of officers beating a man he thought was one of his pals, rather than Mike Cox. Sinsheimer asked whether he saw that officer today. Smut surveyed the crowded gallery looking for Williams, who'd moved back from the first row where he'd sat during Mike's testimony.

But Smut found him. "The guy in the green suit back there," he said.

The case was moving along as planned, but Sinsheimer and Roach were in for an unexpected comeuppance. Roach wanted Craig Jones to testify about the comment Dave Williams made back at the Roxbury station after the beating: "I think my partner hit your partner." Defense lawyers objected, saying having Craig quote Williams was inadmissible hearsay, a legal rule of evidence that generally bars the admission of out-of-court statements made by others. The judge decided he would allow the comment in as an exception to the hearsay rule. But he warned that Roach was at risk of causing a mistrial if he was unable to prove Williams was part of a cover-up by combining this comment with other evidence. "If I let this in and when all the dust settles I don't think it was in furtherance of a conspiracy, this is a mistrial as to Burgio, and maybe as to others.

"Now, do you want it?"

Minutes later, when Roach resumed questioning Craig Jones, he never asked what Dave Williams had told him during a private moment at the station. Mike's lawyers had, in effect, blinked, hoping the rest of their case was enough for the jury to get Burgio.

The jury never heard one of most powerful pieces of evidence.

Jurors were developing a bad feeling about Boston cops, however.

Burgio's longtime partner, Leonard Lilly, for one, left them scratching their heads in wonder when he testified he had no recall of anything Burgio said after the beating. Lilly had driven to Woodruff Way to pick up his pal for a ride back to the station.

"Why did Burgio need a ride?" Sinsheimer asked.

"I don't recall, sir," Lilly said.

"Didn't come up at all?" Sinsheimer asked.

"I didn't say that. I said I don't recall."

"Well," Sinsheimer remarked. His voice turned sarcastic, wanting to draw the jury's attention to the witness's implausible memory failure. "Tell us *everything* you recall about your conversation with Burgio."

"I don't recall any conversation."

Following Lilly, Sergeant Detective Dan Dovidio surprised jurors with his story of driving to the dead end after the beating and seeing only Williams and Burgio.

"Who, if anyone else, was there?" Sinsheimer asked.

"No one," the patrol supervisor said, even though the jury had already learned that after the beating the dead end was filled with cruisers and cops. Dovidio explained the reason he was summoned was that the cruiser operated by one of his charges, Dave Williams, had been damaged, and he was required to inspect it.

"Take your time," instructed Sinsheimer, "and tell the ladies and gentlemen what you did to inspect the vehicle because of the damage."

"I looked at the damage that I could visibly see, and what Officer Williams described to me as to how it happened."

"What did Officer Williams describe for you?"

"I think he said he hit a patch of ice and he skidded," Dovidio said. "Hit the steel post that was in front of the vehicle, and that he had collided with the Lexus."

"He didn't say, Gee, Sarge, you ought to take a look at all the blood on the back of my cruiser?"

"No. No."

"That didn't come up at all?"

The wiry and gray-haired veteran eyed Sinsheimer from behind tinted glasses.

"Nope," he said.

"How about the fact that Michael Cox was injured behind that cruiser, bleeding on the trunk, bleeding from his face, bleeding on the ground; did that come up?"

"No, it did not."

Juror Sharon Schwartz was angered, thinking, "He's so full of shit." In their minds jurors could juxtapose the sergeant's words with a photograph—blown up to poster size—showing swirls of Mike's blood on the trunk. It resembled a painting done by a child. The suburban homemaker was further disgusted to learn about the commendation Dovidio wrote a few days later for Williams and Burgio, an honor that was still part of their personnel records.

"You're a cop and you don't see the blood?" juror Bob McDonough was asking himself. How could Dovidio have not seen that? "He was lying through his teeth."

Later, when the jury was shown a photograph of Jimmy Burgio, McDonough was reminded Jimmy Burgio was the one defendant who'd been a no-show. "I'd thought it was strange he didn't come,"

McDonough said. "It seemed arrogant." McDonough studied the photograph, and his impression of the beefy cop with his goatee only hardened. "I was thinking he looked like a guy you wouldn't like. He looked like a Mafioso."

McDonough commuted by train each day to the trial. Walking from South Station to the courthouse, he found himself growing wary of his surroundings; in particular, of any Boston cops on foot patrol or in a police cruiser. If he spotted a cop, he would cross the street to avoid any close contact. "I just didn't like what I was seeing and hearing in court." He knew he was probably overreacting. "But I just didn't know—I mean so many of these guys were lying—and you hear about people getting to juries.

"Talk about the blue wall of silence," he said. "It's real."

It was different come Kenny's time. Like the jurors, Kenny watched as a confederacy of cops took the stand and hemmed and hawed, straining belief. Most days he sat in the gallery with his older sister's husband, off to the left and behind the table occupied by his attorneys, Willie Davis and Fran Robinson. He wanted his turn.

The moment came ten days before Christmas Day—an unseasonably sunny day in the snowless holiday season. Kenny had not done anything special to prepare—no rehearsals, no coaching from lawyers. His lawyers just wanted Kenny to be himself.

"He was ready," Fran Robinson said. "He had a lot to tell."

The lawyers were confident but for one worry: The judge was taking "judicial notice" of Kenny's perjury conviction. He disclosed to jurors the outcome in Ted Merritt's prosecution earlier in the year—Kenny had been found guilty of lying to a grand jury and obstructing justice when he said he never saw Mike Cox at the fence.

"You have to accept it," the judge said. "You can't question it."

But, the judge continued, the jury was to free to decide how to factor that information into this trial. "We'll hear what Mr. Conley has to say about other things." The jury might view the prior conviction as evidence of a cover-up, as evidence undermining Kenny's credibility, or not at all. "It's entirely up to you."

It all meant that even before Kenny uttered a word he was behind the eight ball. "The jury's being told he's a liar," Fran Robinson said. "How do you defend that?"

Their answer was to let Kenny Conley loose, and once he took the witness stand, Kenny found a groove in Willie Davis's hands. "Are you familiar with the term tunnel vision?" Davis asked Kenny.

"Yes, I am."

"Would you tell the jury what you mean by tunnel vision?"

"Tunnel vision would mean," Kenny said, "that you lock on a particular object and you would basically black out everything else in the surrounding area."

"And is that what happened to you that night?"

"Yes."

Davis drew attention to reasons that someone in Kenny's shoes might "lock on" a suspect that night: First, a fellow police officer had reportedly been shot at Walaikum's, and, second, as far as Kenny knew, Smut Brown was armed. He spoke in a low voice, a courtroom manner he used by design, requiring jurors to practically lean forward in order to listen carefully.

"Were you ever in a position to see whether or not guns were thrown from the Lexus?"

"No," Kenny said.

"Did you believe at that time that the man you saw running toward the fence may have had a gun?"

"Yes," Kenny said.

Davis had Kenny talk expansively about his "activity log" for that night's shift. Kenny had never claimed credit for arresting Smut, and the other side always portrayed the "omission" as proof Kenny was trying to hide his presence at the cul-de-sac; he didn't want a paper trail leading to him. Kenny explained that, officially speaking, he did not arrest Smut. "Arrest means that I would be arresting him, filling out the reports, and following up in court," he said.

"Did you do that with Mr. Brown?" Davis asked.

"No, I did not." He wrote in his log that he'd "assisted" units from Roxbury and Mattapan. "Because we assisted them," Kenny said matter-of-factly. "They had the initial call," he said. Any time he crossed into another police district, he and his partner were backing up other officers who were "handling the reports and the arrests."

From her seat, Fran Robinson liked what she saw. Kenny was open, respectful, and likable. He was effectively debunking the "Southie stereotype" that Ted Merritt had whipped up and exploited at the criminal trial. "Here's this well-spoken guy on the stand, dispelling the stereotype of dumb jock," she said. "Then his rapport with Willie dispelled the other key element of the Southie stereotype—racist."

Most important, they took aim at the portrayal of Kenny as a brick in the blue wall of silence protecting Williams and Burgio because of their ties as fellow cops. Kenny explained that, sure, he knew Dave Williams as a classmate in the police academy, but they were not close friends; he rarely saw him outside work.

What about another son of Southie? "Have you ever asked Mr. Burgio to go out and have a couple of beers?" Willie Davis asked.

"No."

"Has Mr. Burgio ever asked you to go out and have a couple beers?"

"No."

"It's fair to say Mr. Burgio is not a drinking buddy?"

"That's right," Kenny said.

The message was clear: Kenny was not one of them.

"I believe that's all, your honor," Davis said.

When their turn came, the defense lawyers called only three witnesses—municipal police officers who'd been at Woodruff Way. It was, at best, a token effort to open up the scene and suggest munies, rather than their clients, had hit Mike.

Burgio, Williams, and Daley remained silent. They were not about to testify—that would open them up to a lengthy and wide-ranging cross-examination by Mike's lawyers. Besides, Ted Merritt was making cameo appearances in Courtroom 18, seated in the gallery scribbling notes. No way did any of the three want to risk saying something the federal prosecutor could use against them in his criminal probe. "There was a certain grim reaper quality to his presence," said Kenny's lawyer Fran Robinson.

Fourteen days after the trial began—and four days before Christmas—the lawyers delivered closing arguments to the jury. The defense lawyers, one by one, stood and told the jury that Mike's circumstantial case was weak and flimsy. Tom Hoopes said Mike had nothing on Ian Daley. Daley had tried to arrest Mike, he reminded jurors, and was shocked when he realized his mistake. Didn't that show Daley was not one of Mike's beaters? His client, he argued, did no wrong, saw no wrong, said nothing wrong.

"The plaintiffs serve you a banquet," Hoopes said. "They ask you to eat it and come in and deliver a verdict to them." But time and again they called officers whose testimony was questionable and contradictory—officers like Joe Teahan, Gary Ryan, Craig Jones— and Mike himself. "Is the water a little dirty from the witnesses? Does the wine turn in your stomach a little bit from the witnesses?

"Does the apple have a worm in it? Can you trust that kind of food?"

Tom Drechsler told jurors what Jimmy Burgio would have said in his own defense had he testified—that he was too preoccupied arresting Ron "Boogie-Down" Tinsley to have either hit Mike or seen anyone beating Mike. Echoing Hoopes, the lawyer called Mike's case a lot of "finger-pointing" and "rhetoric." No witness had linked his client to the assault. "In the opening statement, I promised you that no one would come forward and say they ever saw James Burgio do anything, and that promise has been fulfilled."

Dan O'Connell trashed Smut Brown. "Okay, Mr. Brown. We know your pedigree." The lawyer recounted Smut's lengthy criminal record. "Yeah, people can change," O'Connell said, "but this leopard didn't change his spots by any stretch." Smut, he noted, had several pending drug-dealing cases to contend with after this trial ended, and that was Smut's likely motivation for testifying against the police officers. "Do you think for any stretch he doesn't hope to get something in nature of relief from those cases? Do you really think he doesn't expect some assistance."

Smut was damaged goods, he said, a liar. "This is the person they offer to you to believe, when he says David Williams hit Michael Cox. I suggest to you it's entirely unworthy of belief." It wasn't even a close call choosing between Dave Williams or Mike Cox's circumstantial case featuring Smut Brown. "Here's a person, David Williams, I suggest, that comported himself in the best traditions of the Boston Police."

Following a recess, Rob Sinsheimer walked to the podium. Throughout the trial he'd lobbied and finally persuaded Steve Roach that, with his experience, he should deliver their final argument. "May it please the court, counsel, Madam Forelady, ladies and gentleman: It is now my solemn duty and obligation to have just a few last words with you on behalf of Michael Cox."

For weeks he'd been jotting notes with the closing in mind. Lines came to him at night, while driving or taking a shower—ways to organize the speech, tie together the pieces of the case. It was what happened to most trial attorneys, living the case day and night.

"You've heard it said that we have the burden of proof," he told the jury. "We don't duck from that at all." His voice felt dry; it was a mixture of his allergies and the importance of the moment. "What I would like to do is explain to you exactly why we believe the evidence you've seen is more than adequate, indeed is abundant for you to find it's more likely than not that these defendants are the ones who did it."

Sinsheimer had pumped himself up. He was feeling that day after day they had methodically gained momentum. He considered one turning point was Sergeant Dan Dovidio's testimony. "It wasn't so much that it was legally relevant, but you felt the atmosphere changing in the room."

Sinsheimer was confident, particularly regarding the brutality charges against Burgio and Williams. In its best light, the circumstantial evidence, combined with Mike's testimony that a white cop kicked him, pointed to Burgio, while the same evidence, combined with Smut Brown's testimony about Dave Williams, showed Williams was in on it too. It was a huge loss not having the jury learn what Williams told Craig Jones at the Roxbury station, but Sinsheimer still thought they had enough.

The case was not nearly as powerful against Conley or Daley. They'd been unable to add much to the judicial notice of Kenny's perjury conviction and were left relying largely on that fact to argue Kenny saw the beating and was both "deliberately indifferent to excessive force" and part of a cover-up.

With Daley, the judge nearly threw out the claim he was a beater, troubled by the thin evidence. But the case was stronger in showing that Daley left Mike after trying to arrest him—"a deliberate indifference to his medical needs." Moreover, Sinsheimer was convinced Daley, if not a beater, saw and knew plenty about the beating, but had shut down and stonewalled justice.

"There's no contest to the notion that this was offensive to human decency," Sinsheimer said as he stood within a few feet of the jury box. He then marched through the evidence, triangulating the short time frame during which the beating occurred with a head count of the suspected officers at the dead end. He stressed Smut Brown had no motive to lie and was therefore credible, despite his criminal record. He revived Joe Teahan's arrival and statement that maybe four or five cars were already there to seal the scene. "I suggest to you if we can account for those four or five cars, then we've met our burden of proof." Then he recapped a process of elimination that zeroed in on his targets.

"Now, what do you hear from the defendants?"

Sinsheimer had reached a key and climactic line, a moment when he'd planned to turn and point outward to the gallery where the defendants sat. It would be a dramatic gesture setting up the answer to the question he'd left hanging in the air.

But as he did, as he looked out, he saw Mike Cox seated alone, and he caught Mike's eye. Mike appeared to be choking up. Sinsheimer was thrown off briefly. "It made me emotional," he said later, "and I thought for a fraction of a second I might lose it, appreciating the magnitude of the case and what Mike had gone through." He turned back to the jury and collected himself. "What do you get in response?" he asked.

"Nothing but lies," he said. "And silence."

Sinsheimer paused to let the words sink in.

"Lies and silence," he repeated.

Sinsheimer walked back to his chair. Judge Young told jurors it was time for them to deliberate. "We should have lunch in there for you now," he said. The judge said they could eat first and then begin their discussions or they could start working right away. "Once we give the case to you, you are in charge."

Everyone in the courtroom stood as the jury left at 1:22 P.M. The judge's chief clerk led them across the hallway behind the courtroom. They entered a rectangular-shaped room with a long oak table in the middle. The jury room's four windows had an unremarkable view of a sprawling asphalt parking lot and, beyond it, the piers and undeveloped waterfront of South Boston. A court officer was stationed outside.

Some jurors grabbed sandwiches and drinks while others used the bathrooms at the far end. Soon enough each had chosen one of the oak chairs padded with blue cushions, and, on the afternoon of December 21, with the holiday season in full swing, they took up the case. "We wanted to get out of there," Bob McDonough said, "but we also wanted to do the right thing." Carol Goslant was the foreperson, and before taking a vote on any of Mike's claims, "we began by discussing the events of the night."

Judge Young, in his instructions, had gone over the six questions the jury was to decide. "Let's come to the first question," he'd said. Question One involved Jimmy Burgio, Dave Williams, and Ian Daley: Did any of them use excessive force? Mike Cox "doesn't complain about being grabbed or stopped, even if the whole thing was a mistake," the judge explained. "What he complains about is the use of excessive force."

Question Two involved only Burgio: Did he commit assault and battery against Mike? The question covered the accusation that Burgio kicked Mike in the face. Question Three also involved only Burgio: Did he intentionally inflict emotional distress on Mike?

Question Four involved all four—Burgio, Williams, Daley, and Kenny Conley: Were they deliberately indifferent to the use of excessive force? Question Five was for Burgio, Williams, and Daley: Were they deliberately indifferent to Mike's medical needs?

Question Six involved just Dave Williams and Kenny Conley: "Question Six is the so-called cover-up theory," the judge said. The legal definition for a cover-up was tricky. Federal civil rights law did not cover an *attempted* cover-up. It only addressed a cover-up that succeeded in blocking justice. So if the jury ruled against the officers in any of the other questions, then that meant the alleged cover-up had failed. "There's no violation for an attempted cover-up that fails." Question Six would be moot.

In sum, the judge advised them: "You apply your common sense to evidence in this case as you are reasonable men and women so that justice may be done."

Now inside the jury room, Carol Goslant had the verdict slip before her. She was in charge of organizing the jury's consideration of each question. "I was eager to get to it," she said later. For the first time in more than two weeks, everyone was free to talk about the charges, the evidence, and the witnesses. It seemed complicated at the start—all the legal principles combined with having four defendants—but very quickly Goslant discovered everyone readily shared her view that Mike's civil rights had been trampled during a brutal assault. The deliberations would revolve around meting out responsibility.

"I was heartened," she said. The jury dug in and began with Question One.

Elsewhere in the courthouse, the waiting game had begun. While the jury deliberated throughout that afternoon and into Tuesday morning, the lawyers mingled in the hall, hung out in the

cafeteria, called into their offices to deal with housekeeping matters for their many other cases and clients. In all his years of practice, Willie Davis never hung around during jury deliberations, but
that was because the old federal courthouse was located only a few
blocks from his office in Faneuil Hall. The new courthouse was
too far a walk, so he stayed put. His partner Fran Robinson, meanwhile, was holed up back at the office preparing for the worst. She
was trying to figure out ways to protect Kenny's assets—as small
as they were—in the event the jury found him liable. The jury had
his perjury conviction as evidence, and that alone could be used to
decide against Kenny. She couldn't bear hanging out in court. "I was
worried," she said. Kenny was doing his best to keep up routines—
errands, the gym, walking his dogs—and Fran had his home and
cell phone numbers on hand to call as soon as there was any news.

Mike was in the courthouse, sticking close to Roach, Sinsheimer,
and Roach's partner and law school classmate Robert Wise. Kimberly was not with Mike; she was working. "I didn't want her there,"
Mike said. He wanted to spare her the ordeal. Despite the stoic exterior, Mike wasn't feeling particularly optimistic. He didn't doubt
the accused cops were guilty, but no one else—not Bob Peabody,
not Ted Merritt—had been able to make a case against them. "No
one had proven it. I didn't expect to."

Rob Sinsheimer, as antsy as he tended to be, resisted talking
about the case. Jury speculation was a useless, wasteful exercise, "a
rookie thing to do—to wonder, guess, prognosticate." He did his
best to busy himself and try not to think about the trial.

Inside the jury room, jurors had made considerable headway
by Tuesday morning. They'd discarded the defense's chaos theory
as unpersuasive. It was a "smokescreen," Carol Goslant said, that
failed to obscure the wrongdoing. "We knew right away these were
the officers who had been involved." Deciding Burgio and Williams
were culpable, she said, turned out to be a "fast decision." That
meant Question Six—the one about a cover-up and the one directed
at Dave Williams and Kenny Conley—was moot. They barely discussed it, except to note its irrelevance based on the other findings.

The sticking points were Ian Daley and Kenny. Some jurors were convinced Daley was one of the beaters while others argued the evidence hadn't shown that. It was a question the jury kept circling back to before deciding not to hold him liable in the beating. Kenny Conley also drew considerable attention. Most had been impressed and found him likable, but Bob McDonough was one who wasn't sure that was enough. "The tunnel vision—I didn't go for that." He wondered whether Kenny was like "the nice young kid next door who's really selling drugs." It was a tough one to figure out.

But other jurors were unwavering about Kenny—that he'd been unfairly entangled in the scandal. "He was honest, straightforward, and his story never changed," Sharon Schwartz said. Listening to his testimony about the foot chase and tunnel vision convinced her Kenny did not see Mike or the beating. "The way he testified solidified my point of view that he was innocent." She couldn't believe he'd been convicted of perjury.

"Most people felt he got a raw deal."

Jurors went round and round—about Daley and, to a lesser extent, Kenny. By Tuesday afternoon, they'd ironed out remaining differences and reached unanimity. Through the court officer sitting outside the room, Goslant sent word to the judge. The jury then hung around waiting to be led back into the courtroom. They'd completed their work, to be sure, but a number felt a bit shortchanged, as if their hands were tied.

"I was frustrated higher-ups weren't on trial," Schwartz said. "I was frustrated this was a civil trial and not a criminal trial." To her, the six questions had not gone far enough. "I would like to have ruled against the whole department." The case, she thought, exposed a dangerous culture of lying that went beyond the four accused officers. "I would like to have said there was corruption in the Boston Police Department."

Mike was downstairs when he heard the verdict was in. He rode the elevator and hurried down the hall to Courtroom 18. The lawyers were gathering at their tables inside the well. Mike took a seat in the gallery toward the back. The news was breaking so quickly he hadn't called Kimberly; even if he had, she wouldn't be able to get to the courthouse in time. The courtroom seemed empty compared to during the trial. Mike realized he practically had the bench to himself. No one was to his left or right. He was alone.

It was 3:45 P.M. when the jury took their seats in the jury box.

"Madam Forelady," Judge Young said, "has the jury reached a unanimous verdict?"

"Yes, we have." Carol Goslant handed the verdict slip to the clerk.

The clerk took the paper. The jurors stood up in the jury box.

"On the use of excessive force theory," the clerk read, beginning with the jury's ruling on Question One, "We find for Michael A. Cox against David C. Williams and James Burgio."

Continuing reading for the jury, the clerk then announced Burgio had committed an assault and battery on Mike. And that Burgio and Williams had been deliberately indifferent to the use of excessive force. And that Ian Daley had been deliberately indifferent to Mike's medical needs.

The clerk's voice carried throughout the courtroom, but the rulings weren't registering with Mike. He thought the clerk was saying the jury had decided for Burgio and Williams, that they were not liable. "I didn't hear it right." Maybe it was because he was so conditioned to bad outcomes. "Nothing had gone well since *that* night."

Mike sat frozen in his seat. But in a few more seconds the truth began to sort itself out, and once that happened, once Mike could feel the truth sinking in and taking hold, he came undone. The clerk was completing the reading of the judgment against Burgio, Williams, and Daley. Mike Cox leaned forward and covered his eyes. He wept.

The judge dismissed the jurors, and the courtroom cleared. In the days to come, Carol Goslant would take her daughter to one of Boston's holiday traditions—a performance of Tchaikovsky's *Nutcracker Suite*. She saw officers directing traffic in the city's theater district and wasn't sure she'd ever think of Boston police in the same way. "It scared me a little." Mike Cox stayed with her. She wrote him a holiday card with a nondenominational greeting: "Peace and Good Will to All." Inside she wrote Mike was a "man of high integrity" and "courage." She wished him well.

Inside the courthouse, word of the verdict had spread quickly, as reporters scrambled to catch up to a breaking story that would be front-page news in the next day's *Boston Globe* and *Boston Herald*.

Willie Davis had immediately called Fran Robinson. "It's okay," he announced happily. "It's okay." Robinson tracked Kenny down. He was working out at a gym in Southie and had heard the good news on the radio. Kenny took some deep breaths. He was still facing prison—this jury's verdict didn't change that. But something felt very different. He'd had his day in court, and when he was given the chance to tell his side, a jury backed him.

Steve Roach, Rob Sinsheimer, and Robert Wise led Mike out of the courtroom and onto the elevator. The mood was restrained. "What for me was a high level of professional achievement was for him, at best, bittersweet," Sinsheimer said. They'd won, but Mike was far from feeling this was "a champagne popping moment." They took the elevator to the second floor, where they got off and began walking down a sweeping stone staircase into the lobby of the grand courthouse. The lawyers planned it that way. Television lights and cameras awaited them. There were more media than at any point during the trial about the worst case of police brutality in the city's history.

The attorneys stepped up to the microphones. "Michael Cox and his family are pleased that the jury saw through the code of silence," Steve Roach said. Roach did most of the talking. Mike made only a few comments. "I just told the truth." His eyes were still wet, and he struggled to gain control of his emotions. "I just told the truth."

Mike was feeling "happy in a sad kind of way." He stepped past the cameras and the frenzied media bustle, and headed for the door. The hour was barely five o'clock, but outside it was twilight. Snow would soon begin to fall, replacing the day's fog and rain. Mike walked out the door and into a city ablaze in holiday lights. He just wanted to get home, where he was Mike Cox, thirty-three, a husband and father of three. But for all the tumble of emotions, one fundamental truth was clear: He'd done it. He'd won justice where everyone else had failed. Leaving the civil rights trial behind, he found himself thinking about the police force he felt had betrayed him, and he said to himself, "Okay, do you believe me now?"

Epilogue

Following the verdict, the spotlight turned from the individual officers to Police Commissioner Paul Evans and Boston Mayor Tom Menino. Based on Judge Young's plan for Mike's civil rights action, a second trial against the department and the city was scheduled to start in early 1999.

The court of public opinion was now fully behind Mike. Leading the charge was Brian McGrory, a popular metro columnist in the *Boston Globe*, who decried the city's determination "to do battle with the decorated cop it failed to help, in a civil trial scheduled to begin next month."

In one column McGrory wrote: "There is a single, significant difference between the brutal beating of Michael Cox in Mattapan by a group of Boston police officers and the savagery committed against Rodney King on a Los Angeles highway in 1991: a videotape.

"Picture the consequences if a neighbor at the end of Woodruff Way stepped outside with a camera in those pre-dawn hours of Jan. 25, 1995, to record a group of cops kicking and pummeling Cox as he rolled into a fetal position on the frozen ground.

"CNN would have replayed the tape at the top of each hour for days, highlighting the horrific twist that the victim was, in fact, a plainclothes Boston police detective mistaken for a suspect. A blue-ribbon commission would have been named to address the brutality within the Boston force. Shouts for sweeping reform would have echoed across the political landscape.

"Instead, the quiet rustling coming from the city government is the sound of Mayor Thomas M. Menino and Commissioner Paul Evans seeking cover from controversy."

Menino and Evans soon changed course. In early February, their lawyers worked out a settlement with Mike's attorneys. There would be no trial—no opportunity for jurors to consider whether Menino and Evans condoned a police culture that turned a blind eye to police brutality and lying. The city would pay Mike and his family about $900,000 plus attorneys' fees, which pushed the package to more than $1.3 million.

Mike was relieved to be done with it. The financial settlement was fine, to be sure. For him, though, the case was never about money. "I'd gladly give every dime and some to go back to how my life was back before this ever happened."

Just before the settlement was made public, Mayor Menino finally uttered his first public comment in four years, telling the *Boston Globe* the Cox case was "an occasion we'll never be proud of. It's not a happy day, a good day, for the Boston Police Department."

In the end, the only justice would be Mike's justice. "We have hit a blue wall," Donald K. Stern, the U.S. attorney in Boston, said on January 27, 2000, at a press conference called to announce he was shutting down Ted Merritt's investigation.

The deadline for bringing criminal charges—known as the statute of limitations—had expired on January 25, the fifth anniversary of the beating. "We do not feel we have enough evidence to charge anyone with the underlying beating," Stern told reporters. "The federal investigation is in an inactive state."

Burgio, Williams, and any others may have been found liable in Mike's civil rights case, but they'd eluded any criminal prosecution.

The blue wall notwithstanding, the federal probe had faltered in a huge rut of its own making—the misguided squeeze play and perjury conviction of Kenny Conley. Merritt had banked the investigation on Kenny—a miscalculation from which the effort to solve Mike's beating never recovered. Stern, however, would have none of it; he stood by his prosecutor and always defended the controversial conviction of Kenny Conley.

Indeed, Ted Merritt was in for some high praise. The very next year, he was one of three federal prosecutors in the Boston office to receive the U.S. Department of Justice's coveted Director's Award for "superior performance" in a high-profile case.

Over at the Boston Police Department, a seemingly wayward cleanup involving key officers in the Cox affair continued. By the end of 1999, Jimmy Burgio, Dave Williams, and Ian Daley were kicked off the force. They were dismissed for a variety of departmental violations—Burgio and Williams for the use of excessive force and for the cover-up of the beating, and Daley for failing to stop the beating and for lying.

The trio vowed to fight and filed labor grievances. Attorney Tom Hoopes, calling Ian Daley's firing a disgrace, insisted his client was a scapegoat. But in a ruling that followed twenty-five days of testimony stretched out over two years, the labor arbitrator concluded otherwise; Ian Daley was a liar, and his termination had been for just cause.

Jimmy Burgio fared no better. Testifying for the first time in any proceeding, he said he'd jumped out of the passenger side of his cruiser, run around the front of the car, and pounced on Ron "Boogie-Down" Tinsley. He said he pointed his gun at the back of Tinsley's head. "I told him if he moved, 'I'll send you to your maker.'" Burgio said he never took his eyes off the suspect, and the two remained frozen until more help came.

His arbitrator did not buy the do-no-evil account, calling Burgio's story, replete with the Hollywood line about sending Boogie-Down to his maker, "implausible" and "demonstrably false," and ruled "substantiated misconduct" warranted his firing.

Burgio went to work as a pipe fitter and moonlighted as a bouncer. Nancy Whiskey's was long closed down, so he took a job working the door at another infamous Southie bar that for decades was controlled by crime boss Whitey Bulger. Back then the bar was Triple O's and nicknamed the "Bucket of Blood" for all the underworld shakedowns and much worse that went on in the back rooms. Following the bust-up of the Bulger gang in the late 1990s, and by the time Jimmy Burgio became regularly employed there, the pub had been renamed the 6 House.

Burgio tried marriage a second time in September 2004, but two years later the marriage was over. He missed being a cop. Burgio did not budge when asked about Woodruff Way: He did not know who beat Mike Cox, but if he had to guess, he said Ian Daley probably had something to do with it, and, of course, as his lawyer always argued, "the munies."

Lastly there was Dave Williams; he appeared before a labor arbitrator who might as well have said, Dave, this is your lucky day. The arbitrator overturned Williams's firing. He ruled Williams deserved only a one-week suspension for filing the false report about him and Burgio being in two cruisers. He said Williams's testimony about chasing Jimmy "Marquis" Evans and knowing nothing about the beating was believable. The arbitrator discounted Mike Cox's testimony as unreliable, dismissed Smut Brown's account entirely, and ignored the jury's verdict in Mike's civil rights trial.

Williams was on the comeback—reinstated and paid an estimated $300,000 in back pay. He went through a refresher course at the police academy in the fall of 2005 and was on the street by early 2006, working as a patrol officer in downtown Boston.

It was a reversal of fortune many familiar with the Cox case found shocking. "A jury of twelve good citizens made a fair decision," said Rob Sinsheimer. "I find it appalling to the point of shocking the conscience that this could be effectively overturned by a single arbitrator. Every citizen who cares about policing and the courts should be outraged."

Smut Brown stood at the gas pump at the Mobil station in Mattapan Square one spring day after the trial when Mike Cox pulled up to another pump to fill up. During their chance encounter, the two found they shared a common ground: Both had long felt harassed by ghosts from the Boston Police Department.

"He told me about getting phone calls in the middle of the night," Smut said later. For his part, Smut complained to Mike that he felt the cops were trying to hunt him down for cooperating. Mike had some counsel for Smut. "I told him that if what he says is true about being targeted, it might be best for him and his family to leave town."

Smut should have heeded Mike's advice, but for reasons that did not involve Boston police. He and eight others in the KOZ gang were indicted in late 1999 on federal drug and conspiracy charges. Smut had been caught twice selling crack cocaine to an undercover agent.

Smut was off the street—for good now, caught up in the nation's draconian mandatory drug sentencing laws. His attorney, Bob Sheketoff, sought out Ted Merritt, hoping to get some leniency for Smut's cooperation, but none was forthcoming. Smut eventually pleaded guilty to one drug sale, and he was in Portland, Maine, in 2003 finishing a seventy-month sentence when he foolishly got caught up in a $300 coke deal. Smut was living at a halfway house and working at a Portland hotel. Instead of heading home to Indira and a third child, a daughter named Destiny, conceived during a conjugal visit and born on May 7, 2004, Smut was sentenced to do another twenty-one years in prison based on sentencing formulas that factored in his multiple prior convictions.

Smut did his best to adjust to prison life. He participated in a Bible study program, took classes in computers and writing, worked out regularly, and became known on the basketball court for his three-point shot. "It's Smiggity for three!" he'd yell after releasing the ball on a long arc to the hoop. In the Portland County jail he became an inmate "trustee" and was known among the guards for his calming presence. A number of the corrections officers wrote on his behalf prior to his final sentencing, including one who credited Smut with saving him from a beating by another inmate. "Mr. Brown saw what was taking place and placed himself in harm's way by assisting me in dealing with the aggressive inmate until my backup arrive," the guard wrote. "I have repeatedly thanked Mr. Brown for his help that day only for him to humbly say, 'I just did what was right.'"

In quiet moments, Smut took to composing dozens and dozens of rap songs, which he'd send home to his mother, Mattie, or to Indira for safekeeping. He dreamed of someday becoming a songwriter. One song he wrote included a verse about his misbegotten role as a witness for Ted Merritt in the conviction of Kenny Conley.

> *Ted Merritt hope you hear it, here's the lyrics:*
> *I can't bear it, tell the public I never lied*
> *And told you that Ken Conley was the white guy.*
> *I'm the star witness Feds want a fast conviction*
> *Exploited my words had an innocent cop sentenced.*
> *So now I'm finished, every cop hatin' my guts*
> *Now my days are dim, I'm condemned for being Smut.*

The year after the trial, Willie Davis told Kenny Conley just before Labor Day to get ready to report to prison. So Kenny and Jen got married that weekend. "It was a forty-eight-hour wedding," Jen said. They rounded up about fifty of their friends and family. Jen's brother flew in on short notice from Chicago. Her aunt drove up from Connecticut. Friends chipped in to put the newlyweds up at a new hotel along the South Boston waterfront, where they spent what Jen later called a "Boston long weekend honeymoon."

The next week, Willie Davis managed to win a stay of Kenny's sentence in order to continue Kenny's legal appeals—a pursuit that turned Herculean. Kenny's circle of supporters expanded to include Boston city councilors and Massachusetts congressmen Joe Moakley and Bill Delahunt. Delahunt called on Bob Bennett, the powerful Washington, D.C., attorney, who represented President Bill Clinton during his sex scandals, for help. Bennett agreed to take over Kenny's appeal on a pro bono basis.

Kenny's case worked its way back and forth between federal court in Boston and the First Circuit Court of Appeals. Ted Merritt and U.S. attorney Donald Stern fought him at every turn. But then in the fall of 2001, Bob Bennett thought he'd achieved a breakthrough. Negotiating with the new U.S. attorney in Boston, appointed by the newly elected President George W. Bush, he'd come up with a deal whereby Kenny would plead nolo contendere, or no contest. Kenny's conviction would stand, but, in return, prosecutors would ask the court to change his sentence to probation: no prison time. Bennett saw it as the kind of "win" he'd worked out in the past for a high-powered clientele of public figures and corporate titans caught up in criminal legal woes that included hard time.

Kenny right away said, No way. The lawyers urged Kenny to sleep on it, talk to Jen. But Ken had the same answer the next morning: No way. "I can't do a no contest," he said. "In my eyes this is a guilty plea, and I wasn't guilty." Taking the deal would also mean he could never return to the police force, "and that's the thing I wanted."

They kept up the fight in court. By now, a centerpiece of the appeal was a document they could thank Bobby Dwan for. During his own legal wrangling after he was put on administrative leave, Bobby devoured documents his attorney secured from the department as part of the discovery process. He came across one that seemed weird—the FBI agent's report about Richie Walker wanting to be hypnotized to better recall Woodruff Way. Bobby realized this had never come out at Kenny's trial and, after he notified Kenny, his attorneys were enraged they'd never seen the FBI memo before.

The lawyers took over from there. The FBI report on Walker, they argued, should have been disclosed prior to Kenny's trial for impeachment purposes. While Bennett raised other issues on appeal, it was the FBI memo that ultimately proved to be the smoking gun needed to finally turn Kenny's conviction on its head. "For me, the sockdolager is the FBI memorandum," wrote the federal judge, employing a fancy word meaning "conclusive blow." The jurist presiding over the appeal happened to be Judge Young, and, on August 14, 2004, Young decided that Kenny had been wrongfully convicted by federal prosecutor Ted Merritt. "Because the government has withheld crucial information, Kenneth M. Conley did not receive a fair trial," he ruled.

When the court of appeals upheld Young's decision the next July, and the U.S. attorney decided to drop the case, Kenny Conley was at last free and clear. When he showed up for the midnight shift on October 4, 2005, at his old station in the South End, the usually perfunctory roll call just before midnight erupted into a welcome back celebration, where Kenny was presented his old badge—number 1016—as a keepsake.

He was thirty-six now, happy to be back in the only job he'd ever wanted. "Knowing me, I'll be a cop until I'm sixty-five." He was promoted to detective in 2007. In a recent interview, he said that while working the streets had changed—more guns and shootings—nothing much had changed about the police force. His colleagues still had the tendency to mess around with the truth regarding an arrest: Too many cops add things to their reports to make their case seem stronger—the very problem people like Bill Bratton called testilying.

"You can't add things," Kenny said one day over coffee. He'd been through a lot because of other cops' lies and truth tailoring. "You either have it or you don't."

"My name is Michael Cox, and I work for the Boston Police Department." Mike spoke those words on December 9, 2000, two years to the day since he'd testified in *Cox et al. v. City of Boston et al.*

But instead of U.S. District Court in Boston, Mike was bearing witness at Harvard Law School. He was one of six panelists addressing the "Impact of Brutality on Victims and Family," part of a daylong symposium entitled "Race, Police and Community." Mike sat in front of the crowded hall dressed in a navy blue blazer, a dark blue shirt, and matching blue tie—a getup that looked like a civilian version of police blue.

He was an oddity—the only cop on the panel; a cop who, strangely enough, was also a victim of police brutality. The other panelists—a Latino woman whose son was killed by police, an activist attorney who'd been involved in countless police misconduct cases—did not hesitate to condemn police while sharing their experiences with police brutality. Seated to Mike's right was the marquee speaker: Abner Louima. The Haitian immigrant had been traveling the country, often in the company of the Reverend Al Sharpton and attorney Johnnie Cochran, talking about his brutal assault and sodomy by New York police officers. "Police brutality happens every day," Louima said. "Unless we all sit down and find a solution, it is going to happen again and again."

Mike followed Louima. He'd sat there stiffly, looking straight ahead, rarely turning sideways to look at Louima or any of the other speakers. "It's tough for me to be here right now and talk about this kind of stuff," he told the audience. "For two reasons: One, it's personal to me, and two, I'm still a police officer." He began by describing generally what happened at Woodruff Way. When the moderator interrupted to ask for specifics, Mike refused. "I think I feel comfortable with just saying [I was] brutalized."

His remarks lasted less than eight minutes. The point he wanted to make, he said, was that the problem of police brutality and the failure to confront it was rooted in "the police culture, and the lack of addressing the police culture by most organizations.

"The culture is just so strong," he said. "Police don't even like to talk about brutality, and ending and stopping it and doing something about it. Many of them don't participate in brutality but they don't realize they are participating by not saying anything."

As the years passed, Mike stayed on at the Boston Police Department. "First, I'm not a quitter," he said by way of explanation. Besides, he'd always felt safer on the inside and had never wavered in his belief that law enforcement was an honorable profession. "I like doing police work," he said. "There are many good things done on a day-to-day basis that you never hear about."

Mike was promoted to deputy superintendent in early 2005 by a new police commissioner, after Paul Evans left the force to take a job with the British government, evaluating the effectiveness of law enforcement agencies in that country. Mike even began taking courses at Suffolk University Law School in Boston.

But nothing would ever be the same, even if Mike felt vindicated by his successful lawsuit. When it came to police work, he was the first to admit this. "It's different for me now," he said. "That trust, I don't have it anymore, no doubt about that."

The change went further, though. "He's not the same guy, not the same cheerful guy," said Mike's best pal from boarding school, Vince Johnson. "I recognize his voice, but that's it. His whole demeanor—there's not that confidence."

Whereas Mike's self-image was once built around cop and career, he now talked about no longer letting the work define him. He focused instead on family and took pride in his kids, watching his sons excel in ice hockey and football as they made their way through high school. The boys were college-bound and still he'd never sat down and talked to them about the night everything changed on January 25, 1995.

Mike was different, he knew that. But the thing he wasn't sure about was whether the police culture was different as a result of his quest for justice. Sure, on the one hand, he'd observed a "heightened awareness of some of the issues."

But the big question was: Could it happen again? Could another police beating like his happen again, where the assailants were shielded afterward by a powerful blue wall of silence?

"I don't know," he said. "I like to think not. But I don't know."

ACKNOWLEDGMENTS

I want to thank the Boston police officers and law enforcement officials who helped me in my research. I want to thank, especially, Mike Cox. He was always a reluctant player in this project. Not surprisingly, Mike never wanted to be in the position of beating victim, and he was not enthusiastic about the case becoming the basis for a book. Given that, he nonetheless did not seek to obstruct the research and, indeed, at various points, sat with me for hours for a number of long interviews. I also want to thank Jim Carnell, Jim Rattigan, Bobby Dwan, Paul Farrahar, Jim Hussey, Jim Burgio, and Craig Jones. Some officers wished to go unnamed. I want to thank Kenny Conley; his wife, Jen; his sister Kris; and their families. The police department's press officer, Elaine Driscoll, and her staff were always gracious and helpful in tracking down public records.

In the Suffolk County District Attorney's office, I want to thank DA Dan Conley, press officers Jake Wark and David Procopio; in the U.S. attorney's office, I thank assistant U.S. attorneys Ted Merritt and Brian Kelly.

I thank Robert Brown, Mattie Brown, Indira Pierce, and their families for their openness and cooperation.

I am grateful to a number of attorneys, some formerly prosecutors, for their assistance: Robert Andrews in Portland, Maine; Saul Pilchen, Jonice Gray Tucker, and Robert Bennett of Washington, D.C.; Robert George, Robert Peabody, Tom Giblin, Robert Sheketoff, Willie Davis, Fran Robinson, and Roberta Golick—all of the Boston area. I would like to acknowledge the assistance of Superior Court Judge Carol Ball and U.S. District Court Judge William G. Young and his staff.

My agent, Richard Abate, made this book a reality. Thanks to Dan Conaway at HarperCollins for seeing the merit in the story, and thanks especially to Gail Winston, my editor, for her unflagging support, not only for this book but for my previous ones as well. I also thank Sarah Whitman-Salkin, Shea O'Rourke, and the entire HarperCollins team.

I was helped in my research with a grant from the Fund for Investigative Journalism, and I completed the book while a Visiting Journalist at the Schuster Institute for Investigative Journalism at Brandeis University. I thank Florence Graves, the institute's executive director, for her continued support for this project.

This book would not have been possible without the support and friendship of my colleagues at Boston University's College of Communication and the *Boston Globe*.

At BU, Bob Zelnick, past chair of the journalism department, and Lou Ureneck, the current chair, helped me juggle my classes with my writing. Mitch Zuckoff has served as a longtime sounding board for ideas (and writing partner), and I thank him. Susan Walker, Paul Schneider, Bill Lord, Ken Holmes, John Schultz, and Sheryl Jackson-Holliday all helped out, with good humor, music, and various other ways. For research assistance, I thank BU journalism graduate students Erin Crosby, Rushmie Kalke, Emily Berry, and John Eagan.

At the *Globe*, past and present, I want to thank Larry Tye, Steve Kurkjian, and Shelley Murphy. Head librarian Lisa Tuite and librarians Richard Pennington and Wanda Joseph-Rollins were invaluable in finding and organizing news stories and photographs. I'd also like to acknowledge Mary Jane Wilkinson, the paper's managing editor for administration, and Toby Leith, the paper's content licensing manager, for their help. Thanks to David Butler for producing a fine map of Roxbury and beyond.

Gerry O'Neill read an early draft of the first half of the book, and I thank him for his insights, his partnership on other books, and his friendship going on two decades now. I've learned so much about journalism and investigative reporting from him. My gratitude goes to another longtime pal, Dave Holahan, who, like Gerry, offered his thumbs-up after an early reading of the work-in-progress. I was lucky to have his wife, Kyn Tolson, and their son, Jackson, also read the manuscript. I thank my parents, John and Nancy Lehr, for their support.

David Bernstein, a writer at the *Boston Phoenix*, was helpful in tracking down stories regarding Jim Burgio that ran in his paper. Carl Todisco, the current owner of 60 Winthrop Street, was generous in giving me a tour of the house where Mike Cox grew up. Thanks also to Donna Kenney, a veteran ER nurse, for a tour of the former Boston City Hospital's emergency room, which was arranged by Maria Pantages and Ellen Berlin of Corporate Communications, Boston University Medical Center. I thank Mike's former classmates at the Wooster School, Vincent Johnson and Tim Fornero, for sharing their memories. Thanks to Rande Styger of the Wooster School for providing yearbook photos and other information about Mike's years at Wooster.

I thank my sons for their help and interest along the way— Christian for his expertise on hip-hop music and Nick for his careful reading of the manuscript. My daughters, Holly and Dana, are, quite simply, pure joy, and they made sure I didn't become an obsessive recluse during the research and writing.

My wife, Karin, was an inspiration throughout, listening to all the stories that are part of a project like this—about the people, the reporting discoveries, the obstacles, the highs, the lows, the breakthroughs. She read each chapter as it was written, with a keen eye, and was indispensable in creating the writing room day-to-day to keep the book moving.

Now that this one is done, it's time for Temenos.

AUTHOR'S NOTE ON SOURCES

Since 1997, I have written a number of in-depth newspaper and magazine articles for the *Boston Globe* about the Cox beating, Mike Cox, Kenny Conley, and Robert "Smut" Brown. In a city where politics, sports, and crime are king, the mistaken and horrific beating of Mike Cox by his fellow cops became a major story, a tragedy that was at once unprecedented and unforgettable. While the brutality may have resulted from a series of misjudgments and mistakes, what happened afterward turned the case into something more—a symbol of an intractable police culture. The failure to bring the beaters to justice became a drama featuring tribalism, abuse of power, race, and policing in a post–affirmative action America. Cities and towns all across the United States confront at times the toxic mix of police brutality and corruption; the Cox case, unlike any other, dramatically illuminates the powerful gravitational pull on a cop to lie. The blue wall of silence held fast even when it meant standing silently as the beating of one brother went unsolved and another was sentenced to serve time in federal prison.

The Fence is a work of nonfiction. The characters are real. No one's name has been changed. The book is based in dozens of interviews with participants and thousands of pages of testimony and other materials that are part of an official record that includes court trials, local and federal investigations, and labor arbitration proceedings. Listed below are those sources. Either by letter or other means, I sought to interview key people involved in the case. Most, but not everyone, cooperated. Fortunately, I was able to draw on the sworn testimony and official statements of those who declined to be interviewed.

The scenes and dialogue are based on the recollection of at least one participant. For grammatical purposes, I occasionally altered the verb tense in a quotation. Where there was conflict regarding an event or someone's words, I was guided by the weight of the evidence—what seemed most plausible and reliable based on the reporting and interviews I conducted, along with the testimony, government records, and court rulings.

NOTES

Please note that full forms are given in the appendices that follow.

PROLOGUE: JANUARY 25, 1995

Sworn testimony and statements by: Donald Caisey (Cox ACU);
  Kimberly Cox (Cox trial deposition, Daley arbitration); Michael
  Cox (Cox IAD, Suffolk GJ, Cox trial deposition, Burgio arbitra-
  tion); Craig Jones (Suffolk GJ, Cox trial deposition, Daley arbi-
  tration); Gary Ryan (Suffolk GJ, Federal GJ, Daley arbitration);
  Isaac Thomas (Cox IAD, Cox ACU, Suffolk GJ, Burgio arbitra-
  tion); Thomas "Joe" Teahan (Daley arbitration); Richard Walker
  (Suffolk GJ, Federal GJ, Burgio arbitration); David Williams
  (Cox ACU, Suffolk GJ, Williams arbitration).

Interview: James Rattigan: Oct. 27, 2005.

Other: Tour of emergency room at Boston City Hospital, Dec. 14,
  2006. Tour and Suffolk County Property and Deed Records for
  52 Supple Road, Dorchester. Tour of Woodruff Way, Mattapan,
  multiple visits.

Records: Audiotape of Boston Police Department radio channel 9,
  Jan. 25, 1995. EMT records of Lyle Jackson, Jan. 25, 1995. Boston
  City Hospital medical records for Michael Cox, Jan. 25, 1995.
  Boston City Hospital medical records for Lyle Jackson, Jan. 25,
  1995; Jan. 31, 1995.

News articles: Chacon, Richard, "Four Men Arrested in Dorchester
  Shooting," *Boston Globe*, Jan. 26, 1995.

CHAPTER 1: MIKE COX

Sworn testimony and statements by: Kimberly Cox (Cox trial deposition); Michael Cox (Cox trial deposition, Cox trial, Dateline

NBC Nov. 2, 1999, Burgio arbitration, Daley arbitration); Craig
Jones (Cox trial deposition).

INTERVIEWS: Michael Cox: March 12, 2006; Oct. 15, 2006. Craig
Jones: Aug. 21, 2007. Seleata Emery: Oct. 10, 2005. Tim Fornero:
Nov. 29, 2005; Dec. 1, 2005. Vincent Johnson: January 2006. Carl
Todisco: July 8, 2005; July 13, 2005.

OTHER: Tour of 60 Winthrop Street, Roxbury, July 8, 2005.

RECORDS: Massachusetts Department of Vital Statistics: birth cer-
tificate of Michael Anthony Cox, no. 9665: June 17, 1965. Wooster
School Yearbook: Class of 1984. Suffolk County Property and
Deed Records for 60 Winthrop St., Roxbury. Boston Landmark
Commission records for 60–62 Winthrop Street, Roxbury. 1962
photograph of 60 Winthrop Street, Roxbury: Bostonian Society,
Robert Severy Collection. Brighton 13 case.

BOOKS AND ARTICLES: Beatty, *The Rascal King*, p. 166. Lukas, *Common
Ground*, p. 17 and after. O'Connor, *Boston A to Z*, p. 27. History of
the Carmelite Monastery, Roxbury: www.carmeliteofboston.org.

NEWS ARTICLES: "New Roxbury Police Division Opens," *Boston Globe*,
March 8, 1971. "Masked Pair Loot Brookline Home of Publishing Executive," *Boston Globe*, Jan. 20, 1974. "Key, 4 Prisoners Disappear from Boston Police Lockup," *Boston Globe*, Jan. 16, 1975.
"3 Policemen Suspended in Escape of 4 Prisoners," *Boston Herald*,
July 8, 1975. Currier, Ann-Mary, "Business Declines as Crime In-
creases in Dudley Square Area of Roxbury," *Boston Globe*, March
27, 1995. Keeley, Bob, "Couple Invade Brookline Home, Seize
Paintings," *Boston Herald*, Jan. 20, 1974. Miller, Margo, "Roxbury's New $3.5 Million Courthouse Opens with Quiet Fanfare,"
Boston Globe, Oct. 27, 1971. Taylor, Jerry, "Police District 2—Leads
City in Violent Crime," *Boston Globe*, June 6, 1977.

CHAPTER 2: ROBERT "SMUT" BROWN

SWORN TESTIMONY AND STATEMENTS BY: Alton Clarke (Boston Police
Department interview, Oct. 18, 1995); Jimmy "Marquis" Evans
(Jackson murder trial); John "Tiny" Evans (Jackson murder trial);
Venice Grant (Boston Police Department interview, Feb. 9, 1995);

Marcello Holliday (Jackson murder grand jury, Feb. 15, 1995); Marvette Neal (Boston Police Department interview, Feb. 10, 1995); Stanley Pittman (Boston Police Department interview, Feb. 20, 1995); Kenneth Renrick (Boston Police Department interview, Feb. 18, 1995; March 9, 1995); April Ross (Boston Police Department interview, Feb. 14, 1995); Marcus Wiggins (Boston Police Department interview, Feb. 14, 1995); Willie Wiggins (Boston Police Department interview, Jan. 25, 1995; May 20, 1996).

INTERVIEWS: Robert Brown III in the Cumberland County Jail, Portland, Maine: March 8, 2006; March 9, 2006; April 3, 2006; July 25, 2006; May 25, 2007. Mattie Brown: April 14, 2006; Sept. 28, 2006; April 18, 2007; Jan. 20, 2008. Indira Pierce: May 7, 2006. Brian Roman: Jan. 13, 2007. Robert Sheketoff: Nov. 13, 1997; Dec. 14, 2005; Dec. 30, 2005; March 17, 2006.

OTHER: Tour and archival photograph research: Cortee's, 324 Washington Street, Dorchester. Tour and archival photograph research: Walaikum's, 451 Blue Hill Avenue, Grove Hall. Tour of apartment number 224, 11 Franklin Hill Avenue, Franklin Hill Housing Project, Boston. Tour of 231 West Selden Street, Mattapan.

RECORDS: Suffolk County Property and Deed Records for 231 West Selden Street, Mattapan. Boston Police Department Arrest Booking Sheets, Jan. 25, 1995 for Robert Brown III, Ronald Tinsley (aka Darryl Greene), Jimmy Evans (aka Robert White), and John Evans (aka Anthony Wilson). Boston Police Department Mugshot Forms, Jan. 25, 1995, for Robert Brown III, Ronald Tinsley, Jimmy Evans, and John Evans. Massachusetts Criminal Offender Record Information (CORI) for Robert Brown III, Ronald Tinsley, Jimmy Evans, and John Evans. Letter by Robert Brown III to U.S. District Court Judge George Z. Singal, District of Maine, July 26, 2006. U.S. Federal Probation Reports on Robert Brown III: April 3, 2001; May 11, 2006. Boston Police Department Homicide Information Sheet: Lyle Jackson. Boston Emergency Medical Transport Report (Jay Weaver), Jan. 25, 1995, for Lyle Jackson. Boston Police Department crime scene photographs of shooting victim Lyle Jackson, Jan. 25, 1995.

NEWS ARTICLES: "Man Stabbed in Roxbury Melee," *Boston Globe*,

Nov. 14, 1995. Barnicle, Mike, "Lyrics Reflect a Bleak Reality,"
Boston Globe, June 6, 1995. Delgado, Luz, "Restoring Neighbor-
hood Pride in Four Corners," *Boston Globe*, Sept. 23, 1993. Saun-
ders, Michael, "How New Kids Reach the End of Their Road,"
Boston Globe, June 9, 1994. Saunders, Michael, "From the Street
to the Fleet, Rap Has Gone Mainstream, but Your Parents May
Not Mind," *Boston Globe*, Dec. 6, 1995. Saunders, Michael, "Super
Jam Puts Hub on Hip-Hop Map," *Boston Globe*, Dec. 9, 1995.

CHAPTER 3: KENNY CONLEY

SWORN TESTIMONY AND STATEMENTS BY: Kenny Conley (Cox trial deposition; Dateline NBC, July 12, 1999); Bobby Dwan (Cox IAD;
Dwan arbitration).

INTERVIEWS: Kenny Conley: July 5, 2001; July 6, 2001; July 11, 2001;
July 16, 2001; Oct. 11, 2005; Nov. 4, 2005; Jan. 19, 2007; Feb. 19,
2007; March 27, 2007. Kristine Conley Cox: Nov. 3, 2005; Dec.
19, 2005; Feb. 19, 2007; several e-mail exchanges. Bobby Dwan:
Dec. 31, 2005; Dec. 5, 2006.

OTHER: Tour of H Street area, South Boston, with Kenny Conley,
Oct. 11, 2005. Tour of 78 H Street, South Boston, with Kristine
Conley Cox, Nov. 3, 2005. Ride-along with Kenny Conley and
Bobby Dwan through South End, Grove Hall, Roxbury, and Mat-
tapan to Woodruff Way, Dec. 5, 2006. Tour of police chase route
from Walaikum's to Woodruff Way, April 21, 2006.

RECORDS: Massachusetts Department of Vital Statistics: death certificate of Maureen Louise Burton Conley, no. 007550: Nov. 24,
1994; death certificate of Kenneth M. Conley, no. 004029: July
20, 2006. Suffolk County Property and Deed Records: Nov. 3,
1977, purchase of 78 H Street, South Boston, by Kenneth and
Maureen Conley, book 9005, p. 198. 2000 U.S. Census information for 78 H Street, South Boston. 1970 and 1980 demographic
data for 78 H Street, Boston Redevelopment Authority. Diary by
Boston Police Lt. Det. John Daley: May 19, 1989, entry.

BOOK: O'Connor, *Boston A to Z*, pp. 295–300.

NEWS ARTICLES: Articles in the *Boston Globe* and *Boston Herald* on the death of Boston police detective Sherman Griffiths, Feb. 17, 1988, through April 1995.

CHAPTER 4: THE TROUBLED BOSTON PD

SWORN TESTIMONY AND STATEMENTS BY: Kenny Conley (Cox trial deposition).

INTERVIEW: Robert Brown III: March 8, 2006; March 9, 2006.

RECORDS: St. Clair, James D. (chairman), St. Clair Report. Shannon, James M. (Massachusetts attorney general), "Report of the Attorney General's Civil Rights Division of Boston Police Department Practices," Dec. 18, 1990. Affidavit of James M. Shannon, former Massachusetts attorney general, on Oct. 15, 1998, in the Cox trial.

COURT CASES: *Commonwealth v. Adams et al.*, 416 Mass. 558 (1993). *Commonwealth v. Adams*, Superior Court Civil 91–3150: Findings of Fact, Conclusions of Law, and Order by Superior Court Judge Hiller B. Zobel, May 8, 1992.

NEWS ARTICLES: "Chronology of the Stuart Case," *Boston Globe*, Jan. 5, 1990. Brelis, Matthew, "In City Newsrooms, Cynics Were Believers: The Stuart Murder Case," *Boston Globe*, Jan. 8, 1990. Craig, Charles, "Witness Tells of Police Beating," *Boston Herald*, Jan. 14, 1992. Craig, Charles, "Handcuffed Driver Was 'Badly Bruised' Says Former Assistant Attorney General," *Boston Herald*, Jan. 23, 1992. Lehr, Dick, "Doubt Cast over Tiffany Moore Verdict," *Boston Globe*, May 4, 2003. Murphy, Sean P., and Locy, Toni, "St. Clair Panel Blasts Police Failings, Urges Roache's Ouster, Other Changes," *Boston Globe*, Jan. 15, 1992. Thomas, Jerry, "Mission Hill Wants Action over Searches," *Boston Globe*, Feb. 9, 1990. Wong, Doris Sue, "Woman Tells of Hearing Wails, Seeing Police Beat Motorist," *Boston Globe*, Jan. 14, 1992. Wong, Doris Sue, "Witness Saw Officers 'Do Nothing Wrong,'" *Boston Globe*, Jan. 23, 1992.

CHAPTER 5: MIKE'S EARLY POLICE CAREER

SWORN TESTIMONY AND STATEMENTS BY: Kimberly Cox (Cox trial deposition); Michael Cox (Cox trial deposition; Dateline NBC,

Nov. 2, 1999); David Williams (Williams arbitration); Richard Walker (Burgio arbitration).

INTERVIEWS: James Rattigan: Oct. 27, 2005. Seleata Emery: Oct. 10, 2005.

RECORDS: Boston Police Department statistics on racial and ethnic makeup, 1988–1991 per a Freedom of Information request. 1970 U.S. Census information for Boston. Psychological assessment of Michael Cox by Dr. Ronald P. Winfield: May 11, 1997.

COURT CASES ON AFFIRMATIVE ACTION AND THE BOSTON POLICE DEPARTMENT: *Castro v. Beecher,* 334 F. Supp. 930 (1971). *Castro v. Beecher,* 459 F.2d 725 (1st Cir. 1972). *Castro v. Beecher,* 365 F. Supp. 655 (D. Mass. 1973). *Paul DeLeo Jr., Thomas Barrett, Michael Conneely, Matthew Hogardt, Brendan Dever, Patrick Rogers, Christopher Carr and Brian Dunford v. City of Boston et al.,* U.S. District Court, District of Massachusetts, civil docket 03–12538.

NEWS ARTICLES: Chivers, C. J., "From Court Order to Reality: A Diverse Boston Police Force," *New York Times,* April 4, 2001. Justice, Glen, "Police Commissioner Spends Night on Duty," *Boston Globe,* July 12, 1993. McGrory, Brian, "Bratton Named to Head Police," *Boston Globe,* June 30, 1993.

CHAPTER 6: CLOSING TIME AT THE CORTEE'S

SWORN TESTIMONY AND STATEMENTS BY: Michael Cox (Daley arbitration; Burgio arbitration); Marcello Holliday (Jackson murder grand jury, Feb. 15, 1995); Craig Jones (Cox trial deposition); Stanley Pittman (Boston Police Department interview, Feb. 20, 1995); Louis Reiter (Cox trial deposition); Gary Ryan (Daley arbitration); Thomas "Joe" Teahan (Daley arbitration); Richard Walker (Daley arbitration); David Williams (Williams arbitration).

INTERVIEWS: James Burgio: Sept. 22, 2006. Michael Cox: March 13, 2006. Robert Brown III: May 25, 2007.

RECORDS: Boston Police Department 1.1 Incident Reports: January 25, 1995: no. 50038785: 324 Washington Street "Shots Fired." Cortee's Lounge, no. 50038593: 324 Washington Street "Property Damage." Cortee's Lounge, no. 50037319: Norwell Street/ Harvard Street. Stop of 1984 Peugeot by Richard Walker. Boston Police Department Internal Affairs records for David Williams in the Ivey and Fernandes cases.

CHAPTER 7: THE MURDER AND THE CHASE AND CHAPTER 8: THE DEAD END

SWORN TESTIMONY AND STATEMENTS BY: Robert Brown III (Jackson murder trial); James Burgio (Burgio arbitration); Donald Caisey (Jones arbitration, Daley arbitration); Kenny Conley (Cox IAD, Federal GJ, Cox trial deposition, Cox trial); Michael Cox (Cox IAD, Cox ACU, Suffolk GJ, Federal GJ, Cox trial deposition, Dateline NBC, Nov. 2, 1999, Burgio arbitration, Daley arbitration); Ian Daley (Daley arbitration); John Evans (Jackson murder trial); Jimmy Evans (Jackson murder trial, Federal GJ); Marcello Holliday (Jackson murder trial); Janet Jackson (Jackson murder trial); Craig Jones (Cox trial deposition); Louis Reiter (Cox trial deposition); Gary Ryan (Daley arbitration); Thomas "Joe" Teahan (Daley arbitration); Isaac Thomas (Burgio arbitration, Williams arbitration, Daley arbitration); Ronald Tinsley (Federal GJ); Richard Walker (Daley arbitration); David Williams (Cox IAD, Cox ACU, Suffolk GJ, Williams arbitration).

INTERVIEWS: Robert Brown III: March 8, 2006; March 9, 2006; April 3, 2006; July 25, 2006. Kenny Conley: July 5, 2001; July 6, 2001; July 11, 2001; July 16, 2001; Oct. 11, 2005; Nov. 4, 2005; Jan. 19, 2007; Feb. 19, 2007; March 27, 2007. Michael Cox: Feb. 19, 2006; March 12, 2006; Oct. 15, 2006; April 6, 2008. Bobby Dwan: Dec. 31, 2005. James Rattigan: Oct. 27, 2005.

OTHER: Ride-along with Kenny Conley and Bobby Dwan retracing route of chase, Dec. 5, 2006. Tour of chase route from Roxbury to Woodruff Way, April 21, 2006.

RECORDS: Boston police interviews with shooting witnesses at Walaikum's. Boston Police Department 1.1 Incident Reports for shooting at Walaikum's, including but not limited to report by Ron Curtis, injury reports by James Rattigan and Mark Freire. Emer-

gency medical technician Jay Weaver's report on the condition
of shooting victim Lyle Jackson. Emergency medical technician
report on the condition of Michael Cox. Boston City Hospital
medical records on the condition and treatment of Michael Cox.
Medical records by private doctors in the ongoing treatment of
Michael Cox following his beating.

NEWS ARTICLES: Associated Press, "Officer Taped Kicking Black Man
Is Suspended," *Boston Globe*, Jan 19, 1995. Associated Press, "Officer Silent in R.I. Kicking Case; Union Blasts Chief for Comments; Mayor Warns of Disciplinary Action," *Boston Globe*, Jan.
20, 1995. Associated Press, "R.I. Officer Defends Kicking Man
in Melee," *Boston Globe*, Jan. 28, 1995. Associated Press, "Providence Police Face Court Suit," *Boston Globe*, March 4, 1995. Associated Press, "Providence, Suspended Police Officer Reach Deal,"
Boston Globe, May 7, 1995. Associated Press, "Officer in Beating
Back in Uniform," *Boston Globe*, May 10, 1995. Dowdy, Zachary R.,
"For Some, Fear Is Constant, Though Boston Crime Dips Figures
Don't Equal Security, Bostonians Say," *Boston Globe*, Jan. 24, 1995.
Dowdy, Zachary R., "Police Censure Three in Fatal Raid; Super-
visor Suspended 30 Days Without Pay," *Boston Globe*, Jan. 27, 1995.

NOTES: My account of the unfolding of the shooting at Walaikum's
differs from that presented by the government during the Lyle
Jackson murder trial in 1996. In arriving at an account that de-
viates from the theory the government presented at the trial, I
have relied on the confessions of Robert Brown III along with
statements given by John and Jimmy Evans, none of which was
available at the time of the trial. This illustrates what I referred
to in my author's note: that, to resolve discrepancies in people's
accounts, I have gone with the weight of the evidence and what,
in my analysis, is most trustworthy. In a second instance, Robert
Brown has testified that during the foot chase, Kenny Conley
yelled at him, "Freeze, Nigger." Kenny Conley has denied using
those words, and, based upon the entire record, Kenny Conley's
denial is credible. The security guard, Charles Bullard, has testified that he alone captured Tiny Evans, an account contradicted
by police officers at the scene including but not limited to Craig

Jones, Joseph Horton, and Bobby Dwan. I credit the police officers' account.

CHAPTER 9: "8-BOY"

SWORN TESTIMONY AND STATEMENTS BY: Robert Brown III (Federal GJ, Conley trial, Cox trial); James Burgio (Burgio arbitration); Donald Caisey (Suffolk GJ, Daley arbitration, Jones arbitration); Kenny Conley (Cox IAD, Federal GJ, Cox trial deposition, Cox trial); Kimberly Cox (Cox trial deposition, Dateline NBC, Nov. 2, 1999); Michael Cox (Cox IAD, Cox ACU, Suffolk GJ, Federal GJ, Cox trial deposition, Cox trial, Dateline NBC, Nov. 2, 1999, Burgio arbitration, Daley arbitration); Ian Daley (Cox IAD, Daley arbitration); Daniel Dovidio (Cox trial deposition, Cox trial); Bobby Dwan (Dwan arbitration); Jimmy Evans (Federal GJ); Craig Jones (Cox IAD, Cox ACU, Suffolk GJ, Cox trial deposition, Daley arbitration); David C. Murphy (Cox ACU, Burgio arbitration, Daley arbitration); Gary Ryan (Cox IAD, Suffolk GJ, Daley arbitration); Michael Stratton (Cox IAD, Cox ACU); Thomas "Joe" Teahan (Cox IAD, Suffolk GJ, Daley arbitration); Isaac Thomas (Burgio arbitration, Williams arbitration, Daley arbitration); Ronald Tinsley (Federal GJ); Richard Walker (Cox IAD, Cox ACU, Suffolk GJ, Federal GJ, Cox trial, Burgio arbitration, Daley arbitration); Marlon Wright (Cox ACU); David Williams (Cox IAD, Cox ACU, Suffolk GJ, Williams arbitration).

INTERVIEWS: Robert Brown III: March 8, 2006; March 9, 2006; April 3, 2006; July 25, 2006. Kenny Conley: July 5, 2001; July 6, 2001; July 11, 2001; July 16, 2001; Oct. 11, 2005; Nov. 4, 2005; Jan. 19, 2007; Feb. 19, 2007; March 27, 2007. Michael Cox: Feb. 19, 2006; March 12, 2006; Oct. 15, 2006; April 6, 2008. Craig Jones: Aug. 21, 2007. James Rattigan: Oct. 27, 2005.

OTHER: Audiotape of Boston Police Department radio channel 3, Jan. 25, 1995.

RECORDS: Boston Police Department 1.1 Incident Reports and Form 26 Reports regarding the events of Jan. 25, 1995, written by Ian Daley, Donald Caisey, Michael Cox, David Williams, James

Burgio, Richard Walker, Kenny Conley, Gary Ryan, Thomas "Joe" Teahan, Daniel Dovidio, David C. Murphy, Isaac Thomas, James Rattigan, Mark Freire, Joseph Freeman, Michael Stratton, Charles Bogues, Margaret A. Waggett, Marlon Wright, David C. McBride, Andrew P. Creed, Richard Houston, Ronald E. Pirrello, John F. McBrien, Ronald Curtiss, Gilbert Alicea, Robert McLaughlin, Ricky F. Cooks, T. Lynn, T. McManus, Daniel Duff, Gregory N. Webb, Keith Webb, John W. Ezekiel, E. Murphy, Leonard J. Lilly, Timothy Brady, Andrell Jones, Bernard A. Doyle, John M. McDonough, Richard Ross, Roy Frederick. Boston Municipal Police officer reports by Andrew Fay, J. Wilmot, David J. O'Connor, Joseph Robles. Sentinel Security incident reports by security officers Paul McGovern, James Waldron. Report by Lt. Kevin D. Foley, commander of the Anti–Gang Violence Unit, to Dep. Superintendent William Johnson, on Jan. 30, 1995: "Injured on Duty: Michael Cox." Boston Police Department disciplinary record for Daniel Dovidio. Boston City Hospital medical records on the condition and treatment of Michael Cox. Physician records on the treatment of Michael Cox following his beating. Psychiatric evaluations of Michael Cox as part of his civil rights case. Emergency medical technician report on Michael Cox.

NEWS ARTICLES: Gelzinis, Peter, column, *Boston Herald*, Jan. 30, 2001. Lehr, Dick/Spotlight Team, "Boston Police Turn Against One of Their Own," *Boston Globe*, Dec. 8, 1997.

NOTES: David Williams has denied making statements attributed to him by Craig Jones; Jones's account is more credible, given the weight of the evidence, which includes the viewpoint of Suffolk County investigators. James Burgio has denied making statements at the Roxbury police station attributed to him by Michael Stratton; Stratton's account is more credible. Ian Daley has denied making statements Donald Caisey and Isaac Thomas have attributed to him; the latter two are more credible: Richard Walker has said he does not recall seeing Craig Jones at the bottom of the hill; Jones's account is corroborated by at least one other officer and is therefore credible.

CHAPTER 10: NO OFFICIAL COMPLAINT

SWORN TESTIMONY AND STATEMENTS BY: Kimberly Cox (Cox trial deposition, Dateline NBC, Nov. 2, 1999); Michael Cox (Cox IAD, Cox ACU, Suffolk GJ, Federal GJ, Cox trial deposition, Cox trial, Dateline NBC, Nov. 2, 1999, Burgio arbitration, Daley arbitration); Luis A. Cruz (Cox trial deposition); James Hussey (Cox trial deposition, Dwan arbitration); Craig Jones (Cox IAD, Cox ACU, Suffolk GJ, Cox trial deposition, Daley arbitration).

INTERVIEWS: Michael Cox: Feb. 19, 2006; March 12, 2006; Oct. 15, 2006; April 6, 2008. Paul Farrahar: June 23, 2006. James Hussey: July 11, 2006; Oct. 1, 2007; several e-mail exchanges. Craig Jones: Aug. 21, 2007. Robert Peabody: Jan. 13, 2006; Jan. 27, 2006; June 26, 2006; Nov. 20, 2007; several e-mail exchanges.

RECORDS: Boston Police Department Form 26 Reports regarding events of Jan. 25, 1995 (see list of officers in Chapter 9 notes). Medical records for Michael Cox. Boston Police Internal Affairs Control Forms for case nos. 2795, 2793, and 2532. Massachusetts bar examination application by Stephen Roach.

NEWS ARTICLES: "Evans Names 89 Being Promoted to Captain, Sergeant and Lieutenant," *Boston Globe*, Feb. 9, 1995. Chacon, Richard, "Boston Police Investigators Seek Cause of Undercover Officer's Injuries," *Boston Globe*, Feb. 4, 1995. Dowdy, Zachary R., "Police Censure Three in Fatal Raid," *Boston Globe*, Jan. 27, 1995. Dowdy, Zachary R., "After Year at Helm, Evans Takes Stock of Policing Record," *Boston Globe*, Feb. 14, 1995. Ford, Beverly, "Highs and Lows Mark Top Cop's First Year," *Boston Herald*, Feb. 14, 1995. Ford, Beverly; Flynn, Sean; and Mulvihill, Maggie, "Alleged Beating of Undercover Cop Probed," *Boston Herald*, Feb. 3, 1995. Gelzinis, Peter, "A Cop Who Gave Her All Remembered for Her Soul," *Boston Herald*, July 20, 1995.

CHAPTER 11: CAN I TALK TO MY LAWYER?

SWORN TESTIMONY AND STATEMENTS BY: Donald Caisey (Cox IAD); Kenny Conley (Cox IAD, Cox trial deposition); Kimberly Cox

(Cox trial deposition); Michael Cox (Cox IAD, Cox trial deposition); Luis A. Cruz (Cox trial deposition); Ian Daley (Cox IAD); Ann Marie Doherty (Cox trial deposition); Bobby Dwan (Cox IAD); Craig Jones (Cox IAD); James Hussey (Cox trial deposition); Isaac Thomas (Cox IAD); Richard Walker (Cox IAD); David Williams (Cox IAD).

INTERVIEWS: Kenny Conley: July 5, 2001; July 6, 2001; July 11, 2001; July 16, 2001; Oct. 11, 2005; Nov. 4, 2005; Jan. 19, 2007; Feb. 19, 2007; March 27, 2007. Michael Cox: Feb. 19, 2006; March 12, 2006; Oct. 15, 2006; April 6, 2008. James Hussey: July 11, 2006; Oct. 1, 2007; several e-mail exchanges. Robert Peabody: Jan. 13, 2006; Jan. 27, 2006; June 26, 2006; Nov. 20, 2007; several e-mail exchanges. Craig Jones: Aug. 21, 2007. James Rattigan: Oct. 27, 2005.

RECORDS: Affidavit by Ann Marie Doherty, chief of Boston Police Department's Bureau of Internal Investigations, on July 11, 1996. Brief filed by attorney Thomas Hoopes on behalf of Ian Daley during labor arbitration against the Boston Police Department. Boston Police Department special order 95–16 on March 15, 1995, on "Use of Non-Lethal Force"; special order 95–17 on March 17, 1995 on "Identification of Plainclothes Officer"; Rule 109: Discipline Procedure; Rule 113: Public Integrity Policy. Boston Police Department records on its Early Intervention System, and Internal Affairs records on David Williams. Psychiatric evaluation of Michael Cox by Dr. Jerome Rogoff on May 4, 1995. Massachusetts bar application of Robert Peabody.

NEWS ARTICLES: Aucoin, Don, "Raising the Stakes: Suffolk County DA Ralph Martin . . . ," *Boston Globe Sunday Magazine*, May 19, 1996. Ellement, John, "DA Mulls Grand Jury in Beating of Officer," *Boston Globe*, June 9, 1995. Ford, Beverly, "Cops in Separate Internal Affairs Probes Transferred to Investigative Unit," *Boston Herald*, March 15, 1995. Ford, Beverly, and Mulvihill, Maggie, "Cops Targeted in Probe of Fellow Officer's Beating," *Boston Herald*, May 5, 1995. Molotsky, Irvin, "Endicott Peabody, 77, Dies; Governor of Massachusetts in 60's," *New York Times*, Dec. 4, 1997. Mulvihill, Maggie, "Cop in Internal Probe Faces Other Allegations," *Boston Herald*, May 9, 1995. Mulvihill, Maggie, "Sources:

Witnesses Saw Beating of Officer," *Boston Herald*, Dec. 15, 1995. Mulvihill, Maggie, and Ford, Beverly, "Anti-Corruption Unit of Hub Police to Probe Beating of Undercover Cop," *Boston Herald*, June 7, 1995.

CHAPTER 12: DAVE, I KNOW YOU KNOW SOMETHING AND CHAPTER 13: COX V. BOSTON POLICE DEPARTMENT

SWORN TESTIMONY AND STATEMENTS BY: William Bratton (talk at Harvard Law School, Nov. 14, 1995); James Burgio (Cox ACU, Burgio arbitration); Kimberly Cox (Cox trial deposition); Michael Cox (Suffolk GJ, Cox trial deposition, Dateline NBC, Nov. 2, 1999, Burgio arbitration); Ian Daley (Cox ACU); Craig Jones (Suffolk GJ, Cox trial deposition); David Williams (Cox ACU, Suffolk GJ, Williams arbitration).

INTERVIEWS: Robert Brown III: March 8, 2006; March 9, 2006; April 3, 2006; July 25, 2006. James Burgio: Sept. 22, 2006. Michael Cox: Feb. 19, 2006. Paul Farrahar: June 23, 2006. Robert Peabody: Jan. 13, 2006; Jan. 27, 2006; June 26, 2006; Nov. 20, 2007; several e-mail exchanges. James Rattigan: Oct. 27, 2005. Robert Sheketoff: Nov. 13, 1997; Dec. 14, 2005; Dec. 30, 2005; March 17, 2006.

RECORDS: Affidavit by Ann Marie Doherty, chief of Boston Police Department's Bureau of Internal Investigations, on July 11, 1996. Boston Police Department Form 26 by Ian Daley regarding his claim of harassment by Craig Jones. Boston Police Department Internal Affairs Division case number 179–94 against David Williams, with ruling by Sgt. Mary L. Crowley. *Cox et al. v. City of Boston et al.*, 95-CV–12729-WGY.

BOOK: Skolnick and Fyfe, *Above the Law*, pp. 7, 89, 92.

NEWS ARTICLES: "Roxbury Site Inaugurated for New Police Headquarters," *Boston Globe*, June 4, 1995. Associated Press, "Providence Police Face Court Suit," *Boston Globe*, March 4, 1995. Associated Press, "Providence, Suspended Police Officer Reach Deal," *Boston Globe*, May 7, 1995. Aucoin, Don, "Raising the Stakes: Suffolk County DA Ralph Martin . . . ," *Boston Globe Sunday Magazine*, May 19, 1996. Chacon, Richard, "Nearly 300 Police Set for First

Night," *Boston Globe*, Dec. 31, 1995. Dowdy, Zachary R., "Minister's Widow Sues City Over Bungled '94 Drug Raid," *Boston Globe*, April 5, 1995. Ellement, John, "Rape Charge Against Officer Is Dismissed," *Boston Globe*, Feb. 28, 1995. Ellement, John, "Detective Files Suits vs. Fellow Officers," *Boston Globe*, Dec. 13, 1995. Flint, Anthony, "Bratton Calls 'Testilying' by Police a Real Concern," *Boston Globe*, Nov. 15, 1995. Ford, Beverly, and Mulvihill, Maggie, "Cox Sues BPD," *Boston Herald*, Dec. 12, 1995. Gelzinis, Peter, "A Cop Who Gave Her All Remembered for Her Soul," *Boston Herald*, July 20, 1995. Jackson, Derrick, "Mayor Scrooge," *Boston Globe*, April 7, 1995. Lehr, Dick, "Getting Away with Murder: Charlestown's Code of Silence Means Most Neighborhood Killings Go Unsolved," *Boston Globe*, Oct. 11, 1992. Rakowsky, Judy, "Three Convicted in 'Code of Silence' Trial," *Boston Globe*, March 23, 1995. Zuckoff, Mitchell, "Corruption Probe Shakes Up Boston Detective Unit," *Boston Globe*, Feb. 10, 1996.

NOTES: David Williams has given conflicting testimony regarding his encounter with Mike Cox in Franklin Park. During his own labor arbitration, Williams rejected Cox's version of events; however, during his earlier testimony before the Suffolk County grand jury he confirmed the thrust of Cox's account and testified that during their encounter he defended James Burgio in connection with the Cox beating.

CHAPTER 14: THE WHITE GUY AT THE FENCE

SWORN TESTIMONY AND STATEMENTS BY: Michael Cox (Cox trial deposition); Thomas Dowd (Dwan arbitration); Craig Jones (Cox trial deposition); William Johnston (Dateline NBC, Nov. 2, 1999); David Williams (Williams arbitration).

INTERVIEWS: Carol S. Ball: Sept. 6, 2007; several e-mail exchanges. Robert Brown III in the Cumberland County Jail, Portland, Maine: March 8, 2006; March 9, 2006; April 3, 2006; July 25, 2006; May 25, 2007. Mattie Brown: April 14, 2006; Sept. 28, 2006; April 18, 2007; Jan. 20, 2008. Kenny Conley: July 5, 2001; July 6, 2001; July 11, 2001; July 16, 2001; Oct. 11, 2005; Nov. 4, 2005; Jan.

19, 2007; Feb. 19, 2007; March 27, 2007. Jennifer Conley: July 11, 2001. Michael Cox: Feb. 19, 2006. Robert Peabody: Jan. 13, 2006; Jan. 27, 2006; June 26, 2006; Nov. 20, 2007; several e-mail exchanges. Paul Farrahar: June 23, 2006. Robert Sheketoff: Nov. 13, 1997; Dec. 14, 2005; Dec. 30, 2005; March 17, 2006.

RECORDS: Boston Police Department Internal Affairs Records in the excessive force case against David Williams by complainant Valdir Fernandes (records include a 2007 Freedom of Information request for information regarding disciplinary action taken against Williams, to which the department responded that no record of any disciplinary action could be found). The police officer acquitted at trial in March 1995 of charges of stealing money from a wallet in the Brighton police station was later fired by the department. Trial transcripts from the Lyle Jackson murder trial. Massachusetts bar application by S. Theodore Merritt. FBI Form 302, dated March 31, 1997, by agent Kimberly McAllister.

NEWS ARTICLES: "Guilty Plea in Bomb Threat; City Man Could Face 12-Year Term," *Worcester Telegram & Gazette*, Dec. 18, 1996. Doherty, William F., "Ex-Police Duo Get Three Years for Thefts; Seized $200,000 in Bogus Raids," *Boston Globe*, May 22, 1998. Ellement, John, "Officer Is Acquitted in Theft of $6,350 from Wallet at Station," *Boston Globe*, March 26, 1995. Estes, Andrea, "Electrician Cleared of Death in Fire Sparked by Wiring," *Boston Herald*, Jan. 28, 1997. Nealon, Patricia, "Guard Gets Three Years for Scalding Prisoner," *Boston Globe*, Feb. 21, 1997. Rakowsky, Judy, "Ex-Winthrop Chief Says He Took Bribes, Evaded Taxes," *Boston Globe*, July 26, 1995. Tatz, Dennis, "State Trooper Admits Beating Man After Incident in Bar," *Patriot Ledger*, Sept. 5, 1996.

CHAPTER 15: THE PERJURY TRAP AND
CHAPTER 16: A FEDERAL MISCARRIAGE OF JUSTICE

SWORN TESTIMONY AND STATEMENTS BY: Michael Cox (Cox trial deposition); Robert Brown III (Federal GJ); Ann Marie Doherty (Cox trial deposition); Richard Walker (Cox IAD, Cox ACU, Federal GJ).

INTERVIEWS: Robert Brown III in the Cumberland County Jail, Portland, Maine: March 8, 2006; March 9, 2006; April 3, 2006; July 25, 2006; May 25, 2007. Mattie Brown: April 14, 2006; Sept. 28, 2006; April 18, 2007; Jan. 20, 2008. James Burgio: Sept. 22, 2006. Kenny Conley: July 5, 2001; July 6, 2001; July 11, 2001; July 16, 2001; Oct. 11, 2005; Nov. 4, 2005; Jan. 19, 2007; Feb. 19, 2007; March 27, 2007. Jennifer Conley: July 11, 2001. Kris Conley Cox: several e-mail exchanges. Michael Cox: Feb. 19, 2006; March 12, 2006; Oct. 15, 2006; April 6, 2008. Willie Davis: March 30, 2006. Bobby Dwan: Dec. 31, 2005; Dec. 5, 2006. Ted Merritt: Aug. 22, 2001. Indira Pierce: May 7, 2006. Fran Robinson: March 30, 2006; several e-mail exchanges. Robert Sheketoff: Nov. 13, 1997; Dec. 14, 2005; Dec. 30, 2005; March 17, 2006. Prof. Daniel Simons: July 17, 2001; July 20, 2001.

OTHER: Simons and Chabris, "Gorillas in Our Midst."

RECORDS: Conley trial transcript. Massachusetts bar application of Willie J. Davis. Affidavit by Fran Robinson regarding the arrest of Kenny Conley at FBI headquarters in Boston. Multiple appellate briefs filed by Saul Pilchin and Robert Bennett on behalf of Kenny Conley. Multiple briefs filed by Ted Merritt and other federal prosecutors in connection with the Conley case. FBI Form 302, dated March 31, 1997, by agent Kimberly McAllister regarding her interview with Kenny Conley. FBI Form 302 report by agent Kimberly McAllister regarding her dealings with Richard Walker. Multiple judicial rulings in the Conley case. July 29, 1997, letter by Ted Merritt to Willie Davis regarding indictment of Kenny Conley. Copies of the tickets and fliers for the Ken Conley time at the Florian Hall. Suffolk County Property and Deed Records: property information for the Cox home at Rundel Park in Dorchester.

NEWS ARTICLES: "Narcotics Agents Nab Three in Boston," *Boston Globe*, Jan. 16, 1970. Carroll, Matt, "Witness Says Officer Nearby During Police Beating," *Boston Globe*, June 6, 1998. Carroll, Matt. "Boston Officer Is Convicted of Lying in Police Beating," *Boston Globe*, June 11, 1998. Cavaan, Azell Murphy, "Cop Found Guilty of Perjury," *Boston Herald*, June 11, 1998. Doherty, William F.,

"Willie Davis Resigns as US Magistrate, to Defend Cleaver," *Boston Globe*, Jan. 25, 1976. Ford, Beverly, "Boston Officer Sues Fellow Cop for Alleged Perjury over Beating," *Boston Herald*, Feb. 4, 1998. Gladwell, Malcolm, "Wrong Turn: How the Fight to Make America's Highways Safer Went off Course," *New Yorker*, June 11, 2001. Lehr, Dick, "The Case for Kenny Conley," *Boston Globe Sunday Magazine*, Sept. 23, 2001. Lehr, Dick, "Witness in '95 Brutality Case Offers New Account," *Boston Globe*, Sept. 17, 2006. Ranalli, Ralph, "Officer Sentenced for Perjury in Cop-Beating Case," *Boston Herald*, Sept. 30, 1998.

NOTES: Several additional points about Smut Brown seeing the "tall, white cop" from the fence in the hallway of federal court during the Conley trial. First, FBI agent Kimberly McAllister disputes the account given by Smut, his mother, and Indira Pierce. In several telephone conversations with McAllister and a Boston FBI spokeswoman, McAllister initially denied ever having spoken with Smut Brown in the hallway during a recess in the trial. She later amended her position, saying she recalled an encounter but that Smut Brown never pointed out someone else (not Kenny Conley) as the man standing next to the Cox beating at the fence. Had Brown done that, she said, she immediately would have notified federal prosecutors.

Second, Brown's account of never having seen Kenny Conley before that day in court surfaced during interviews for this book. Brown's disclosures debunked the accepted lore of the Conley trial that he had indeed provided an in-court identification, and many lawyers involved in the case—attorneys for Mike Cox and Kenny Conley—were incredulous; in large part, I believe, because to accept Brown's statements meant admitting they had all overlooked a major flaw in the government's case; namely, that Brown, contrary to appearances, had in fact never identified Kenny Conley. But Brown's full story holds up when examined closely, and he certainly had no incentive or motivation to lie. During the book interviews he was facing sentencing in a federal drug case, and his account could only potentially hurt him by angering federal officials.

Finally, some leading members of the Boston bar were astounded that Brown's effort to alert the FBI about a potential witness/suspect in the beating went unheeded. "This would be explosively important evidence in favor of the defendant [Conley]," said J. W. Carney Jr. "Any federal agent who received this information from a critical witness and intentionally suppressed it by not informing the federal prosecutor is, in my opinion, guilty of obstruction of justice." Robert A. George, another prominent attorney, said, "That's a huge, huge error." He said, "There's a duty, and not just a legal duty, but a human duty, to put a stop to the trial at that point." Both said Brown's statements should have triggered an investigation, but no known inquiry into possible federal misconduct resulted from the new disclosures.

CHAPTER 17: ON HIS OWN

SWORN TESTIMONY AND STATEMENTS BY: Kenny Conley (Cox trial deposition, Dateline NBC, Nov. 2, 1999); Kimberly Cox (Cox trial deposition); Michael Cox (Cox trial deposition, Dateline NBC, Nov. 2, 1999); James Hussey (Cox trial deposition).

INTERVIEWS: James Burgio: Sept. 22, 2006. Kenny Conley: July 16, 2001. Willie Davis: March 30, 2006. Robert Sinsheimer: February 2008 (several). Robert Brown III in the Cumberland County Jail, Portland, Maine: March 8, 2006; March 9, 2006; April 3, 2006; July 25, 2006; May 25, 2007. Mattie Brown: April 14, 2006; Sept. 28, 2006; April 18, 2007; Jan. 20, 2008. Fran Robinson: March 30, 2006; Dec. 10, 2007. Judge William G. Young: April 6, 2005.

OTHER: Tour of 5 Sutton Street, Dorchester.

RECORDS: Cox trial transcript. Letter dated Aug. 24, 1998, from the attorney for Chris Ruggerio to the Boston Police Department regarding the conduct of James Burgio. *Ruggerio v. City of Boston, James Burgio, et al.*, 00-CV-12320-RCL, U.S. District Court, Boston. Massachusetts bar application of Robert Sinsheimer. Affidavit of Ann Marie Doherty, dated June 11, 1996. Psychiatric evaluation of Michael Cox by Dr. Ronald P. Winfield, dated May 11, 1997. Op-ed article written by Boston police commissioner

Paul V. Evans, *Boston Globe*, June 27, 1996. Editorial, "A Criminal Police Silence," critical of Commissioner Evans in the *Boston Globe*, Oct. 10, 1998.

NEWS ARTICLES: Bernstein, David, "$50 million Worth of Mistakes," *Boston Phoenix*, Sept. 28, 2004. Dannen, Laura; Liesener, Katie; and Lux, Rachel, "System to Stem Police Perjury Not Implemented," *Boston Globe*, Oct. 24, 2005. Doherty, William F., "Internal Affairs Head Gets Transfer: Police Unit Was Target of Criticism," *Boston Globe*, Feb. 7, 1998. Farmelant, Scott, "Crossing the Blue Line— 'Code of Silence' May Be Key in Police Abuse Suit," *Boston Herald*, Nov. 22, 1998. Gelzinis, Peter, "Two Hub Cops Get What They Didn't Deserve," *Boston Herald*, Oct. 1, 1998. Murphy, Shelley, "Four Are Disciplined in Beating of Officer," *Boston Globe*, Oct. 29, 1998. Murphy, Shelley, "Convicted Cop's Bitterness Grows; Urges Peers to Own Up to Cox Beating," *Boston Globe*, Oct. 29, 1998. Murphy, Shelley, "Beaten Officer's Lawsuit Merits Trial, Judge Finds," *Boston Globe*, Nov. 25, 1998. Vasquez, Daniel, "Witness of Officer's Beating Is in Hospital After Shooting," *Boston Globe*, Sept. 17, 1998. Vasquez, Daniel, "Beaten Officer Says His Dignity Was Taken," *Boston Globe*, Oct. 30, 1998. Weber, David, "Cop Fired in Beating Case Accused in Bar Assault," *Boston Herald*, Oct. 3, 2000. Zuckoff, Mitchell, "Judge Assails Police, Grants Man New Trial," *Boston Globe*, Dec. 23, 1997.

CHAPTER 18: THE TRIAL

INTERVIEWS: Kenny Conley: July 5, 2001; July 6, 2001; July 11, 2001; July 16, 2001; Oct. 11, 2005; Nov. 4, 2005; Jan. 19, ,2007; Feb. 19, 2007; March 27, 2007. Michael Cox: April 6, 2008. Carol Goslant: March 2, 2008; March 9, 2008. Willie Davis: March 30, 2006; March 4, 2008. Robert McDonough: March 4, 2008. Fran Robinson: March 30, 2006; Dec. 10, 2007; March 3, 2008; several e-mail exchanges. Robert Sinsheimer: February 2008 (several); several e-mail exchanges. Sharon Schwartz: March 7, 2008.

OTHER: Tour of Judge Young's courtroom, lobby, and jury room, April 6, 2005. Q&A with Judge Young in "Bench Conference"

1996 column in *Lawyer's Weekly.* Holiday greeting card dated Dec. 23, 1998, sent by Carol Goslant to Michael Cox.

RECORDS: Cox trial transcript.

NEWS ARTICLES: Campbell, Robert, "A Mixed Verdict on New US Courthouse," *Boston Globe*, Sept. 7, 1998. Donlan, Ann E., "Many Cops Relieved That Justice Is Finally Done," *Boston Herald*, Dec. 23, 1998. MacQuarrie, Brian, "Two Found Liable in Beating of Fellow Officer; a Third Policeman Also Implicated; Another Cleared," *Boston Globe*, Dec. 23, 1998. Murphy, Shelley, "At Civil Rights Trial, Detective Recalls He Was Beaten," *Boston Globe*, Dec. 9, 1998. Ranalli, Ralph, "He Kicked Me in the Mouth; Cox Testifies in Police Beating Lawsuit," *Boston Herald*, Dec. 9, 1998. Ranalli, Ralph, "Lawyer for Accused Cops Attacks Victim's Memory," *Boston Herald*, Dec. 10, 1998. Ranalli, Ralph, "Hub Cops Found Liable in Officer's Beating," *Boston Herald*, Dec. 23, 1998.

EPILOGUE

SWORN TESTIMONY AND STATEMENTS BY: James Burgio (Burgio arbitration); Michael Cox (Dateline NBC, Nov. 2, 1999).

INTERVIEWS: James Burgio: Sept. 22, 2006. Kenny Conley: July 11, 2001; Jan. 19, 2007. Michael Cox: March 12, 2006; Oct. 15, 2006; April 6, 2008. Craig Jones: Aug. 20, 2007. Vincent Johnson: 2006 (several telephone interviews). Bobby Dwan: Dec. 31, 2005; Dec. 5, 2006. Robert Sheketoff: Nov. 13, 1997; Dec. 14, 2005; Dec. 30, 2005; March 17, 2006. Robert Sinsheimer: April 1, 2008.

RECORDS: Boston Police Department rank and promotion records for Isaac Thomas, David C. Murphy, Daniel D. Dovidio, Richard Walker. Videotape and written results of polygraph test administered to Bobby Dwan. Dwan arbitration, Jones arbitration, Burgio arbitration, and Williams arbitration. *United States v. Robert Brown III, Sylvester Gendraw, Maurice Payne, Donald Cook*, U.S. District Court, District of Massachusetts, criminal docket 99–10383. *United States v. Robert Brown III*, U.S. District Court, District of Maine, criminal docket 04–12. Letters written in 2006 by Peter Moulton and other correctional officers at the

Cumberland County Jail in Portland, Maine, on behalf of Robert Brown III. Ruling in 2005 by U.S. District Court Judge William G. Young overturning Kenny Conley's conviction, along with the First Circuit Court of Appeals ruling upholding the judge's decision and the U.S. attorney's decision not to retry Conley on perjury and obstruction of justice charges. Videotape of Harvard Law School forum on Dec. 9, 2000, on police brutality. Videotape of Jan. 27, 2007, gathering at the Florian Hall in Dorchester of friends and family of Kenny Conley.

BOOK: Bennett, *In the Ring.*

NEWS ARTICLES: Cramer, Maria, "Minority Police Group Cites Race in Disciplinary Action: The Murphy Case Raises Questions," *Boston Globe*, Jan. 11, 2008. Dabilis, Andy, "City, Officer Seen Near Settlement in Beating Suit," *Boston Globe*, Feb. 9, 1999. Editorial titled "Batterers with Badges," *Boston Globe*, Dec. 30, 2007. Gelzinis, Peter, "Cop's Suspension Appears to Retire Issues of Beating," *Boston Herald*, Jan. 30, 2001. Latour, Francie, "Two Officers Fined in '95 Cox Beating," *Boston Globe*, Dec. 4, 1999. Latour, Francie, "Criminal Case in Cox Saga Is Closed; No Charges Are Filed in Officers' Beating," *Boston Globe*, Jan. 28, 2000. Latour, Francie, "Sergeant Suspended for Role in Cox Beating Probe," *Boston Globe*, Jan. 30, 2001. Martinez, Jose, "Evans Cans Two Cops Accused in Cox Case," *Boston Herald*, Dec. 4, 1999. Martinez, Jose, "'Blue Wall' Stymies Cop-Beating Probe," *Boston Herald*, Jan. 28, 2000. McGrory, Brian, "One Disgrace After Another," *Boston Globe*, Jan. 12, 1999. McGrory, Brian, "Cox Case Gets More Shameful," *Boston Globe*, Feb. 2, 1999. Murphy, Shelley, "Beaten Officer Still Sees Honor in Police Work," *Boston Globe*, Feb. 12, 1999. Nealon, Patricia, "City Reportedly to Pay Cox $900,000," *Boston Globe*, Feb. 11, 1999. Smalley, Suzanne, "Hub Police Lieutenant Accused of Assault," *Boston Globe*, March 8, 2007. Vasquez, Daniel, "Police Officers Being Scapegoated in Beating of Colleague, Lawyers Say," *Boston Globe*, Jan. 29, 1999. Watson, Jamal E., "Louima, Others, Decry Police Brutality," *Boston Globe*, Dec. 10, 2000.

NOTES: The second "time" for Kenny Conley at the Florian Hall in January 2007 was unabashedly celebratory, unlike the first that was held soon after Kenny's indictment. It was a time for Kenny to thank his supporters. Those who spoke included Massachusetts Congressman William Delahunt and Kenny's Washington lawyers Robert Bennett, Saul Pilchin, and Jonice Gray Tucker.

Following are updates on other police officers.

The three sergeants who'd botched the scene at Woodruff Way all went their separate ways. Isaac Thomas, for one, left the gang unit and went to work in the police station on Blue Hill Avenue in Mattapan. Daniel J. Dovidio and David C. Murphy eventually ran afoul of either department regulations or the law.

Dovidio, despite his obvious malfeasance at Woodruff Way, held on until 2001 when he reached the mandatory retirement age of sixty-five. Even so, a thirty-one-year career ended in disgrace when the department finally got around to holding him accountable. He spent his final days serving a forty-five-day suspension for neglect of duty—for ordering Williams and Burgio to file false reports and for failing to investigate the scene properly when he insisted he never saw Mike's blood on the trunk of a cruiser. The lies, noted the department, "interfered with the ongoing investigation" into the Cox beating.

Murphy made lieutenant but then had his own difficulties. He was charged in April 2007 with assaulting his girlfriend in a bar in Baltimore. The couple had gone to the James Joyce Pub after watching the Red Sox play the Orioles in Camden Yards. Murphy "arched his right hand and punched the white female in the face," according to the Baltimore police report. He fled to his nearby hotel, where he was arrested, police said. One month later, the Boston police lieutenant pleaded guilty in a Baltimore court to punching his girlfriend. He was sentenced to eighteen months of probation.

It turned out not to be Murphy's first arrest for domestic violence. The prior fall he'd been charged with assaulting his girlfriend in their home in Quincy, Massachusetts, but the case was dismissed when the victim did not want to pursue criminal charges.

For his troubles, Murphy barely skipped a beat at the Boston

Police Department. He was given a thirty-day suspension, of which he served only five days.

No question the police department has worked in mysterious ways. Richie Walker, for all of his verbal contortions, never faced any scrutiny or second looks by the department about the discrepancies in his testimony. He went back to work, was promoted to detective, and was reassigned from the station in Mattapan to the one in Hyde Park. In March 2006, he was off-duty at a nightclub near Blue Hill Avenue when he heard gunshots outside. Walker ran out and saw a gunman running in his direction. They traded gunfire before Walker captured the man. It turned out the gunman had just shot two people. Walker received a departmental award for bravery.

Meanwhile, Kenny's partner, Bobby Dwan, who, like Kenny, had consistently cooperated with investigators but was nonetheless swept up in Ted Merritt's investigation, waged a fifteen-month fight to regain his place on the force. He passed a polygraph examination immediately after his suspension in October 1998, hoping to show police officials he had nothing to hide. Commissioner Evans initially was unimpressed, but by fall 1999, Bobby got Evans to agree to honor the results of a second exam. If Bobby flunked, the result could be used against him in departmental hearings. If he passed, he would be reinstated.

Bobby took the second polygraph on January 14, 2000. The examiner was a retired air force lieutenant colonel who'd administered numerous polygraph exams in the military and afterward. He asked questions chosen by police officials. Bobby passed with flying colors. Two weeks later, he returned to active duty. His beef with the department wasn't entirely over, however. He still had to fight the Evans administration to reclaim income lost while he was on leave. But by 2008, his career seemed back on track. Widely regarded as a tough, honest cop, he worked for a while in the Anti-Corruption Unit. He was then promoted to lieutenant and assigned to oversee the night shift in Roxbury.

Then there were Craig Jones's travails. Several years after the jury's verdict in Mike's case, he suddenly found himself under investigation for his punching of Tiny Evans at Woodruff Way. The question was whether Craig had used excessive or justifiable force. The department on January 11, 2002—seven years after Mike's beating—decided against him. He was suspended for nine months. Flummoxed, he appealed, and the next year a labor arbitrator overturned the suspension. The arbitrator criticized the "extraordinary passage of time" in bringing the charges against Craig and then ruled "Jones was neither overzealous nor inappropriate in his exercise of force." Noting Craig had no prior history of excessive force, she ruled there was no evidence that Craig "gratuitously punched Tiny Evans for no reason at all. Whether it was one punch or two, the evidence establishes Jones was reacting strictly to the suspect's continued resistance to arrest even after he was handcuffed." Craig may have won his arbitration, but his career path seemed to have hit a serious snag. He was moved out of the elite gang unit and assigned to a "traffic car" downtown. Craig missed the action of his former post.

APPENDIX A

COURT CASES

Michael P. Carney v. City of Springfield et al., 403 Mass. 604 (1988).

Castro v. Beecher, 334 F. Supp. 930 (1971).

Castro v. Beecher, 459 F.2d 725 (1st Cir. 1972).

Castro v. Beecher, 365 F. Supp. 655 (D. Mass. 1973).

Commonwealth v. Thomas Adams et al., Suffolk Superior Court, Massachusetts, civil docket 89–0046 ("Brighton 13 case").

John L. Smith v. Thomas Adams et al., Suffolk Superior Court, Massachusetts, civil docket 91–3150.

Commonwealth v. Adams et al., 416 Mass. 558 (1993).

Commonwealth v. Robert Brown III, Dorchester District Court, criminal docket 93–1823.

Commonwealth v. Jimmy Evans, Suffolk Superior Court, Massachusetts, criminal docket 95–10237 ("Jackson murder trial").

Commonwealth v. Robert Brown, Suffolk Superior Court, Massachusetts, criminal docket 95–10238 ("Jackson murder trial").

Commonwealth v. Ronald Tinsley, Suffolk Superior Court, Massachusetts, criminal docket 95–10239 ("Jackson murder trial").

Commonwealth v. John Evans, Suffolk Superior Court, Massachusetts, criminal docket 95–10190 ("Jackson murder trial").

Cox et al. v. City of Boston et al., U.S. District Court, District of Massachusetts, civil docket 95–12729 ("Cox trial").

Commonwealth v. Robert Brown III, Suffolk Superior Court, Massachusetts, criminal docket 97–11423.

United States v. Kenneth M. Conley, U.S. District Court, District of Massachusetts, criminal docket 97–10213 ("Conley trial").

United States v. Conley, 186 F.3d 7 (1st Cir. 1999).

United States v. Conley, 103 F. Supp. 2d 45 (D. Mass. 2000).

United States v. Conley, 249 F. 3d 38 (1st Cir. 2001).

United States v. Conley, 164 F. Supp. 2d 216 (D. Mass. 2001).

United States v. Conley, 323 F. 3d 7 (1st Cir. 2003) (en banc).

United States v. Conley, 332 F. Supp. 2d 302 (D. Mass. 2004).

Conley v. United States (1st Cir. 2005) (en banc).

Cox et al. v. Conley et al., U.S. District Court, District of Massachusetts, civil docket 98–10129.

Cox et al. v. Evans, U.S. District Court, District of Massachusetts, civil docket 98–10768.

United States v. Robert Brown III, Sylvester Gendraw, Maurice Payne, Donald Cook, U.S. District Court, District of Massachusetts, criminal docket 99–10383.

Dwan v. City of Boston, 329 F. 3d 275 (1st Cir. 2003).

Commonwealth v. Evans, 786 N.E. 2d 375 (Mass. 2003).

Paul DeLeo Jr., Thomas Barrett, Michael Conneely, Matthew Hogardt, Brendan Dever, Patrick Rogers, Christopher Carr, and Brian Dunford v. City of Boston et al., U.S. District Court, District of Massachusetts, civil docket 03–12538.

United States v. Robert Brown III, U.S. District Court, District of Maine, criminal docket 04–12.

Evans v. Verdini, U.S. District Court, District of Massachusetts, civil docket 04–10323.

Evans v. Verdini, 1st Circuit Court of Appeals, Massachusetts, civil docket 05–2272.

APPENDIX B

BOOKS

Beatty, Jack. *The Rascal King: The Life and Times of James Michael Curley (1874–1958)*. Reading, Mass.: Addison-Wesley, 1992.

Bennett, Robert S. *In the Ring: The Trials of a Washington Lawyer*. New York: Crown, 2008.

Gladwell, Malcolm. *Blink: The Power of Thinking Without Thinking*. New York: Little, Brown, 2005.

Harr, Jonathan. *A Civil Action*. New York: Random House, 1995.

Johnson, Marilynn S. *Street Justice: A History of Police Violence in New York City*. Boston: Beacon Press, 2003.

Lane, Roger. *Policing the City: Boston 1822–1885*. Cambridge, Mass.: Harvard University Press, 1967.

Lukas, J. Anthony. *Common Ground: A Turbulent Decade in the Lives of Three American Families*. New York: Knopf, 1985.

O'Connor, Thomas H. *Boston A to Z*. Cambridge, Mass.: Harvard University Press, 2000.

Skolnick, Jerome H., and Fyfe, James J. *Above the Law: Police and the Excessive Use of Force*. New York: Free Press, 1993.

ARTICLES AND SPECIAL REPORTS

Fisher, Stanley Z. (Boston University law professor). "'Just the Facts, Ma'am': Lying and the Omission of Exculpatory Evidence in Police Reports." *New England Law Review* 28: 1 (Fall 1993).

Gladwell, Malcolm. "Wrong Turn: How the Fight to Make America's Highways Safer Went off Course." *New Yorker*, June 11, 2001.

Shannon, James M. (Massachusetts attorney general); Jonas, Stephen A. (chief, Public Protection Bureau, Massachusetts attorney general's office); Heins, Marjorie (chief, Civil Rights Division, Massachusetts attorney general's office). "Report of the Attorney General's Civil Rights Division of Boston Police Department Practices." Dec. 18, 1990.

Simons, Daniel J., and Chabris, Christopher F. "Gorillas in Our Midst: Sustained Inattentional Blindness for Dynamic Events." *Perception* 28: 1059 (1999).

St. Clair, James D. (chairman). "Report of the Boston Police Department Management Review Committee ("St. Clair Report")." submitted to Mayor Raymond L. Flynn, Jan. 14, 1992.

APPENDIX C

BOSTON POLICE DEPARTMENT RULES AND REGULATIONS

Rule 109: Discipline Procedure
Rule 113: Public Integrity Policy
Rule 301: Pursuit Driving
Rule 304: Use of Non-Lethal Force

BOSTON POLICE DEPARTMENT INTERNAL INVESTIGATIONS

Anti-Corruption Division: ACD 95–012 (Cox beating: "Cox ACU").

Internal Affairs: Complaint Control 2797 (Cox beating: "Cox IAD").

Internal Affairs: Case 383–92. Physical Abuse: P.O. David C. Williams. Not Sustained.

Internal Affairs: Case 080–93. Physical Abuse: P.O. David C. Williams. Exonerated.

Internal Affairs: Case 179–94. Physical Abuse: P.O. David C. Williams. Exonerated.

Internal Affairs: Case 260–94. Physical Abuse: P.O. David C. Williams. Sustained.

Internal Affairs: Case 079–97. Violation of Rights: P.O. David C. Williams. Unfounded.

Internal Affairs: Case 141–97. Violation of Rule 304: P.O. David C. Williams. Not Sustained.

Internal Affairs: Case 312–99. Multiple Rule Violations. Sgt. Det. Daniel J. Dovidio. Sustained.

BOSTON POLICE DEPARTMENT LABOR ARBITRATION PROCEEDINGS

GRIEVANT: Robert Dwan, suspension.

AMERICAN ARBITRATION ASSOCIATION: In the Matter of Arbitration Between the Boston Police Superior Officers Federation & the City of Boston. Case No.: 3900015599

Philip Dunn, arbitrator. Decision: March 27, 2003 ("Dwan arbitration").

GRIEVANT: James Burgio, discharge.

AMERICAN ARBITRATION ASSOCIATION: In the Matter of Arbitration Between the Boston Police Patrolmen's Association & the City of Boston. Case No.: 16–1370.

Tammy Brynie, arbitrator. Decision: April 17, 2003 ("Burgio arbitration").

GRIEVANT: Craig Jones, suspension.

AMERICAN ARBITRATION ASSOCIATION: In the Matter of Arbitration Between the Boston Police Patrolmen's Association & the City of Boston. Case No.: 16–1534.

Roberta Golick, arbitrator. Decision: Sept. 5, 2003 ("Jones arbitration").

GRIEVANT: David Williams, discharge.

AMERICAN ARBITRATION ASSOCIATION: In the Matter of Arbitration Between the Boston Police Patrolmen's Association & the City of Boston.

Lawrence T. Holden Jr., arbitrator. Decision: June 30, 2005 ("Williams arbitration").

GRIEVANT: Ian Daley, discharge.

AMERICAN ARBITRATION ASSOCIATION: In the Matter of Arbitration Between the Boston Police Patrolmen's Association & the City of Boston. Case No.: 16–1393.

Roberta Golick, arbitrator. Decision: Aug. 9, 2005 ("Daley arbitration").

SUFFOLK COUNTY DISTRICT ATTORNEY'S OFFICE

Special Grand Jury: John Doe Investigation (Cox beating). SUCR95–
    11315.
Robert Peabody, assistant district attorney ("Suffolk GJ").

UNITED STATES ATTORNEY'S OFFICE, DISTRICT OF MASSACHUSETTS

Federal Grand Jury: *United States v. John Doe* (Cox beating).
S. Theodore Merritt, assistant U.S. attorney ("Federal GJ").